Farming and the Countryside

Farming and the Countryside

M.H.R. SOPER, OBE, MA, MIBiol, FRAgS, FCGI

E.S. CARTER, CBE, BSc, FIBiol, FRASE, FRAgS

Farming Press

First published 1985
by the Association of Agriculture
as *Modern Farming and the Countryside*
Second Edition 1991

A catalogue record for this book is available
from the British Library

ISBN 0 85236 225 0

**Published by Farming Press Books
4 Friars Courtyard, 30-32 Princes Street
Ipswich IP1 1RJ, United Kingdom**

Distributed in North America by Diamond Farm Enterprises
Box 537, Alexandria Bay, NY 13607, USA

Phototypeset by Typestylers, Ipswich
Printed and bound in Great Britain by Butler & Tanner, Frome, Somerset

Contents

*The colour section **AWARD-WINNING EXAMPLES**
OF FARMING AND COUNTRYSIDE CONSERVATION*
appears between pages 116 and 117

Preface to the First Edition

There has never been a time when there has been such widespread public interest in the countryside, or more questioning of the methods used by farmers in the production of food. Food shortages in two World Wars and the economic situation of Britain today are overwhelming reasons for maintaining an efficient and up-to-date farming industry. Even though there are now food surpluses of some commodities in the Western World, the burgeoning populations and chronic food shortages — even famine — in many developing countries must preclude any serious reduction in food production in Britain or Western Europe.

But on this small island, with its population of some 57 million people, the countryside where the food is produced, diminished annually by 12,000-15,000 hectares of land for housing and industry, must also provide the facilities for recreation and for the maintenance of its natural biological assets and its scenic beauty.

But is it possible to achieve all of these apparently conflicting objectives? In this booklet, we have attempted to analyse the issues, starting with the pressures on the land and the degree of intensity with which farming will need to be practised in the future. This is followed by an account of modern farming and the underlying reasons for the methods used. We then discuss the requirements of the public for recreation, and for the conservation of the natural resources and finally, with the aid of short case studies, we have shown that it is possible for modern farming, conservation and recreation to live together, given the cooperation and the goodwill of all the parties involved in the management of the countryside.

M.H.R. Soper
E.S. Carter
February 1985

Preface to the Second Edition

When the first edition of this book was published in 1985, food *surpluses* in the European Community were becoming a cause for concern, as also were chronic food *shortages* in many parts of the world. Though food stocks in the EC were considerably reduced for a time owing to poor harvests, the potential for large surpluses still remains, with serious economic consequences for farming. Many developing countries still face famine through natural causes or war and, for the fourth consecutive year, world coarse grain consumption will exceed production.

In Eastern Europe, political change has brought about local food shortages, but such change may eventually enable these countries to realise their undoubted agricultural potential. Future GATT decisions and proposals to reform the CAP pose a grave threat to the profitability of farming and the maintenance of the countryside in the UK. In the longer term, the possibility of climatic change could create very different conditions for farming worldwide. Against all these uncertainties, it would be extremely unwise to damage the future agricultural potential of the UK, though it seems that considerable areas of land may have to be taken out of production in the shorter term.

The great majority of farmers have always cared for the land and the countryside, and, since the first edition of this book, they have significantly increased their concern for its conservation. By 1990, for example, FWAG advisers had visited by request no fewer than 20,000 farmers to advise on wildlife and landscape conservation, and its integration with profitable farming. Government initiatives have also encouraged a change in attitude through grants for tree planting and farm woodlands, encouragement for diversification and the establishment of Environmentally Sensitive Areas, which are all designed to help shift the emphasis towards conservation integrated with good farming practice.

Advances in technology are certain to lead to many changes in rural areas and in the use of land. First quality land is likely to remain in food production, but alternative uses may have to be found for land in less favoured areas. Such uses must provide employment for the rural population if the social fabric of the countryside is to be maintained. Cooperation and goodwill between all concerned will be essential if farming, conservation and recreation are to continue to live and develop together.

M.H.R. Soper
E.S. Carter
May 1991

Foreword

BY SIR DEREK BARBER
Formerly Chairman of the Countryside Commission

These days when so many books on farming and the countryside are written by geographers, landscape architects, zoologists and ecologists, what a pleasure it is to read a publication whose authors have had a life-long experience in the science and practice of agriculture and are real countrymen.

This, the second edition of their book, has been largely rewritten where necessary in the light of current developments, but it remains the solid, non-tedious review of the farming and countryside scene which was so much appreciated when it first appeared. It is a wholly *professional* document, yet its chief value rests on the fact that it has been stitched together to be palatable and easily understood by the lay reader who enjoys the countryside but has little or no knowledge of the farming industry.

The whole picture is here: from a review of the pressures on farming and farmland, through a particularly helpful section on modern farming systems, to the practical ways and means of adopting conservation principles. Both farmers and interested urban dwellers should actually *enjoy* reading what these skilled and knowledgeable authors have to say.

It is an extraordinary paradox that in the 1990s there is more interest in farming and its impact on landscapes and wildlife habitats than ever before. Yet all too often it is the articulate siren voices of the prejudiced with their sweeping criticisms of the farming industry who are listened to by those seeking information. Much of the prejudice is based on totally unscientific grounds but because of the media attraction for a "good story", the sensational and dramatic — and lamentably misinformed — report and discussion of some incident is the one we all read in our newspapers.

In the section on 'Modern Production Methods and the Environment' is a wealth of good, solid information and comment to refurbish the minds and attitudes of those who have been harkening to the wrong messages about pollution, soil health and influences on the food produced. Perhaps there may come a realisation to the reader that agrochemicals and sprays to control unpleasant crop diseases are as important to growing the morning's breakfast cereal as petrol is to their motor cars.

But although I see this book as having most benefit for the lay person and particularly for schools, it is vastly interesting and informative for the practising farmer and student. It is all so well condensed with an economy of words married to good presentation.

Throughout, the same emphasis on balance and perspective conveys a real sense of authority. This is an accurate and perceptive account of where modern farming finds itself. This is farming at work. It is also conservation at work. Those in each camp and those who, happily, are in both at the same time will gather much from the labours of these two authors. I commend what they have done with considerable enthusiasm.

1991

Pressures on the Land

THE IMPACT OF URBAN PRESSURE

In a country such as Britain it is impossible to consider farming and the countryside in isolation. Seventy-five per cent of the population, or some forty million people, live in towns and cities of ten thousand or more inhabitants, and such concentrations inevitably have an impact on the rural areas which surround them. But it is not only in the adjacent areas where the impact is felt. In the age of rapid motorised travel, the effect is spread much further afield to areas of the countryside which have particular recreational attractions to those who live in towns. If living standards continue to improve, thus bringing the possibility of travel to many more people, it is unlikely that this pressure will diminish in the years ahead, especially in those areas with potential for tourism. The impact may not be quite as great as was once feared, but it will be there all the same. (See Recreation and the Countryside, page 120).

In addition, further shifts in the population, from larger cities into smaller towns and villages in the fringe areas, will take place and this is likely to affect the countryside in three ways. Firstly, through the loss of rural land taken for housing, light industry, transport and recreation. Secondly, through a certain amount of physical interference with farming in those regions bordering urban developments. Thirdly, through a change in the attitudes of people who move out of the larger towns to live in the country, and who have different ideas as to what should, or should not, take place in the countryside in which they have chosen to live. Greater pressures will certainly be exerted on farmers to conserve natural resources. More concern can also be expected over some farming practices which, to an urban-orientated public with views formed mainly by the media, might seem to be undesirable in one way or another.

Farmers will not be able to isolate themselves, however much they may wish to do so, from these pressures exerted on their

1

own doorstep. Nor ought they attempt to do so, for people are likely to become more insistent that they have a right to express an opinion on how taxpayers' money is spent in supporting the production of their food. If farmers ignored such views, they could run the risk of further reductions in support at a time of surpluses; or the imposition of more restrictive legislation to control farming practices considered unacceptable by the general public. The classic example of what can happen is the restriction on straw burning, which will become illegal after the harvest of 1992. While it is true that this ban was imposed following protests from rural communities, it would probably not have happened so soon if the non-indigenous population in rural areas had not increased so rapidly in the past thirty years.

Any consideration of the impact on the rural environment of the loss of agricultural land involves not only the size of area required for other purposes and its location in the national context, but also the effects of each type of development on the management of the countryside as a whole. Such effects might relate to the provision of additional land for recreation, to the preservation of special amenities, or to demands that farmers should abandon systems of crop or animal production which cause disquiet on aesthetic, amenity or ethical grounds. There have been cases recently, for example, where farmers have had to abandon the keeping of pigs in close proximity to residential areas. Trends such as this are already apparent, and have to be taken very seriously by the farming community if a harmonious relationship is to be maintained between town and country — something which is essential to the well-being of both parties in a densely populated country such as Britain. Already, some extremists on both sides have taken up entrenched positions, and no time should be lost in trying to reconcile the more important of the conflicting interests.

In any analysis of the future of farming and the countryside, the first requirement must be to review the existing position with regard to the actual loss of land from agriculture, the uses to which such land is put, and the likely rate of loss in the years ahead. Having considered these issues it will then be necessary to assess whether the fears often expressed by environmental-ists that Britain could eventually become little more than a vast town with only small areas of farmland left between conurbations, are justified. It is also important to consider whether such loss of land is really serious or not. It may be that annual loss is only small relative to the total area available, and also that less land is now needed for agriculture because food surpluses exist in the European Community and the United States of America.

It then becomes important to establish whether a modern farming industry is justified in the light of those surpluses. If so, the industry's role in relation to conservation of the natural resources of the countryside, either from an ecological, environmental or aesthetic point of view should be considered. That in itself will not be easy. Quite serious conflicts already exist between some of the interests most concerned with conservation topics. It is no longer just a question of balancing modern farming with the preservation of the environment, but one of balancing the demands of recreation and public use with a range of other factors, often of a scientific nature, related to the maintenance of a stable environment. A typical example is the destruction of natural habitats by excessive pressure from tourism. The preservation of an acceptable countryside is therefore a far more complex problem than one of just controlling farming practices.

THE LOSS OF RURAL LAND

Britain is often regarded as the most urbanised country is Western Europe. On a land utilisation basis, however, this is not the case, since only some 11-12 per cent of the land area is actually devoted to urban use, whereas in the Netherlands, it is over 16 per cent. The figure for Britain is lower because of its high proportion of upland pasture and rough grazing compared with neighbouring European countries, and because much of the population tends to be concentrated in large cities and conurbations. It is true that the overall population density is high if measured in terms of land per head — only 0.4 hectare per head. But this is a global figure, and there is great variation over the country as a whole. For example, one has only to think of the average number of people living on a hectare of land in

London, compared with those on a hectare of land in Snowdonia, to appreciate that global averages can mean very little in a local context.

Robin Best in *Land Use and Living Space (1981)*, the most comprehensive study on the subject, classified urban areas according to population size and density, and the different uses to which land was put, such as housing, education, transport, recreation and so on. In 1961, the most up-to-date figures available for his study, the total area falling within the category of urban land was only 1.5 million hectares (4.25 million acres), representing 9.9 per cent of the area of England and Wales, and only 2.6 per cent of the area of Scotland. Since that time, some 14,000-15,000 hectares of land have been lost from the agricultural sector each year, which, if it all went to urban development, would bring the total to approximately 1.9 million hectares, or some 12.5 per cent of the total land area. In fact, this would be an overestimate, since quite considerable areas have not gone into urban use at all; forestry, for example, accounted for a significant acreage in the late 1960s and early 1970s. It can, therefore, be safely assumed that the figure for land in urban use in the UK today does not yet exceed the figure of 12.5 per cent given above.

Regional differences, however, may vary between 5 per cent and 25 per cent of land in urban use, with the South East and the Midlands having the highest proportion, and the far West and the North the lowest. Robin Best defined a relatively narrow band running diagonally across the country from Lancashire/West Yorkshire through the Midlands, down to London and the South East, which contained an above average proportion of land in the urban category. It is in this belt that there has been most development since his base point of 1961. It is in this belt, too, that problems associated with the allocation of land for development are most acute, and the pressure on Green Belts most intense.

Reliable figures for the actual movement of land from agriculture into strictly urban use are difficult to obtain, since not all land which ceases to be recorded in the annual agricultural statistics goes for building houses, factories and roads, though normally a high proportion will do so. In the past, as mentioned, considerable areas were planted with trees, mostly by the Forestry Commission, but more recently by commercial companies, particularly in Scotland and Wales. There may also be quite a long time lag between land ceasing to be recorded in agricultural statistics and being developed.

Statements are made from time to time that areas the size of an English county are taken every ten years for urban development and industrial use. This must be an exaggeration, for if it were true, Britain would be far more heavily urbanised than it is, and the area devoted to such uses would be far more than the 12-13 per cent estimated above.

Overall, housing is by far the most important element in urban land use. Industry, on the whole, is not a very large consumer of land, accounting for only 7-8 per cent of the built-up area, while education in the form of schools, colleges and recreation areas is of a similar order. Naturally, these figures vary to some extent between cities, large towns, small towns and villages. What became clear from Best's analyses, and this still holds true today, is that it is basically the state of the domestic housing industry which has the greatest bearing on the movement of land out of agriculture. There is no reason to suppose that the situation will change materially in the future, though the proportion of land which moves out of food production into recreational use, such as golf courses, water sports, and country parks, may very well increase.

THE PAST THIRTY YEARS

In the 1960s, the loss of land from agriculture appeared to have averaged out at some 16,000-17,000 hectares a year. By the early 1970s, however, the post-war housing boom was over, and most of the new towns were virtually completed as far as the construction of houses was concerned. The use of land for that purpose declined relative to that required for light industry, mineral extraction and road building, which were all consuming quite considerable areas of land. The motorway programme at its peak needed a lot of land for the roads themselves and also for the minerals required for their construction.

Shifts in population were also taking place, with movement of families from cities and large towns into the fringe areas and smaller towns and villages. This shift did not lead to as large a loss of purely agricultural land as might have been expected, since post-war planning regulations effectively prevented the worst excesses of pre-war 'ribbon development'. Very considerable scope also remained for filling in the undeveloped land (infilling) in smaller towns and villages within commuting range of the larger towns and cities. At the time, this saved a lot of farmland from being swallowed up in indiscriminate building, such as would have undoubtedly occurred prior to 1939. The replanning of town centres for offices and shopping precincts reduced inner city population densities to some extent, though the rise in the population due to the post-war bulge prevented these from falling as much as might have been desired. This was the era of high-rise building when an attempt was made to avoid lateral sprawl by building upwards in order to save land, which, in retrospect, proved to be a socially disastrous experiment.

After 1975, the situation changed, mainly due to the state of the national economy. A tight money policy, high interest rates and pressures on Local Authority spending, all combined to reduce public and private building. In addition, the price of land soared, and inflation, induced largely by the 1973 oil crisis, added enormously to costs. The recession that followed also tended to reduce the scale of industrial building, though there was still activity in the provision of light industrial estates in smaller and medium sized towns. There was, too, a reduction in the demand for motorway land, the road building programme having by then passed its peak. This reduction in building activity had further repercussions on the extractive industries through lower requirements for gravel, sand, cement and bricks. So the loss of land from agriculture slowed down for a period of some ten years up to the middle of the 1980s. But even during this period, the pattern was by no means uniform across the country. The South East continued to dominate industrial development, and to attract new residents, in spite of efforts to move industry further to the North.

After 1985, a more buoyant economy reversed, to some extent, the downward trend in the Midlands and the South, and there was a resurgence of building activity. In-filling proceeded apace, and there was a considerable expansion of towns not restricted by Green Belt constraints. The advent of a number of foreign companies stimulated industrial development in the North in order to avoid the shortage of labour in the South and South East. Closer links with the European Community also increased demand for development in the Eastern Counties, now the most rapidly developing area in Britain. There was, and still is, severe pressure on the Government and on planners to release more land for building both private houses and industrial premises, especially around big cities, where restrictions are most severe. Relaxation of Green Belts has so far been strongly resisted by the Government, but the pressure for new building in some areas is now so great that it seems inevitable that some concessions will have to be made (see page 11). From the early 1980s, some effort was put into the regeneration of certain inner city areas with the conversion of old urban industrial sites into housing, shopping and recreational developments — something demanded for years, but which, for various reasons, lagged far behind developments on the extremities of built-up areas, where costs are much lower. The Dockland development in London is a typical example of such a conversion, even though it is one that has attracted considerable criticism.

The building boom of the mid-to-late 1980s was short lived. Growth in inflation and the rapid increase in interest rates had a severe dampening effect on the housing market, and this led once more to a slowing down in building activity to levels similar to those at the start of the decade.

FUTURE DEMANDS FOR LAND

It will already be clear that many different factors can affect the movement of land out of the agricultural sector. This makes the forecasting of future trends very hazardous, particularly as overall national figures can have little relevance to local situations. Forecasting in relation to agriculture is now especially difficult for two reasons. Firstly, because of the surpluses of some temperate foodstuffs which steadily built up in the Community throughout the 1980s, and led to drastic efforts to curb agricultural production through measures such as quotas, Set-aside and price reduction. Secondly, because of the opening up of Britain to the full force of the projected free market in 1992, and, closely allied to it, the completion of the Channel Tunnel in 1993. In theory, this should make it easier for Britain to trade its agricultural commodities in Europe, but it could also leave her at the mercy of a flood of imports from better organised marketing systems in countries such as Holland and France.

The Influence of Surpluses

Another imponderable is the annual increase in production which can be anticipated from the application of new agricultural techniques. This is estimated to have averaged out at some 2 per cent per annum over the past decade. Whether it will continue at this level if the profitability of farming declines, as seems inevitable in the light of surpluses and a period of economic restraint, is somewhat doubtful. But at this point still another factor could come into play. This is the impact of developments in biotechnology, which are forecast to have a dramatic effect on agricultural production in the future. Latest estimates, however, suggest that it may be ten to fifteen years before biotechnology begins to make a major impact on production and agricultural output. So it might be safer to ignore this factor in forecasting changes up to the beginning of the twenty-first century.

But there is no doubt that the most difficult factor in determining future trends in agriculture is the impact of food surpluses, which built up after the first edition of this book was published in 1985. By 1990, surpluses had again been markedly reduced in the case of cereals and milk products, due to poorer harvests, dry summers and financial constraints, and became little more than those needed as a strategic food reserve. But another series of good harvests in Europe, such as occurred in the mid 1980s, would rapidly lead to a build up of unwanted surpluses once more. So the previous emphasis on preserving land from exploitation so that it could remain available for food production, has been completely altered to one of preserving it for the protection of the environment as a whole, and of finding suitable uses for land that may no longer be needed for agriculture, certainly in the short term.

Nevertheless, some attempts at estimating future demands for land utilisation must be made if sensible decisions about the future of the farming industry are to be taken. This has special relevance to the level of intensity at which farming will need to be practised if a high proportion of the nation's food is to be produced at home, and if, in so doing, large and costly surpluses are to be avoided. If considerable areas of land are to be transferred out of agriculture for building and industrial development of various kinds, then there need be less concern about surpluses, and much of the land that is left will have to be farmed fairly intensively. However, if only relatively small areas are to be removed annually, then there will be land surplus to food production requirements, and measures such as the Set-aside scheme, the planting of woodlands, and so on will become increasingly important. In that case, it should be easier to integrate future farming methods with conservation, and some of the more difficult problems of intensification would be avoided.

Varying and conflicting estimates have been made about the amount of land that will need to be taken out of agriculture if unmanageable surpluses of food are to be avoided; for example, that 5.5 million hectares will need to be lost by the year 2015. Such estimates were made at a time when surpluses of cereals, milk, milk products and beef were particularly high. But the two relatively poor harvests in Western Europe in 1987 and 1988, and the more moderate harvest in 1989 reduced them very considerably. This proves how quickly situations can change, and how unwise it would be, in an era of rapidly increasing world population, to cut back production too far, and to adopt policies which would remove land *permanently* from its potential for food production. It should also not be forgotten that in the present century Britain was twice within a few weeks of desperate wartime food shortages, and that although war in Europe may now seem inconceivable, an island nation should always retain sufficient land on which to grow its own food, in case of emergency. Even if that land is not kept permanently in production, it should be available for rapid conversion, if necessary. This concept becomes of even greater significance in the light of the possible consequences of climatic change on world food production.

The Land Required

So how much land is likely to be required for development in the coming years, and how much will be needed for agriculture? It seems unlikely that the levels of house building reached in the post-war years will ever be achieved again in the future, even though the restrictions imposed on the release of land for building have led to a pent up demand for new housing. The high cost of land and of construction seem likely to limit the scale of building, even if restrictions on the release of land were to be removed completely. An important consideration is whether the land required for a modest expansion in domestic housing is to come from the renewal of town and city centres and from in-filling, or whether much of it will come from so-called green field sites.

There is unquestionably considerable scope for urban renewal, but even if this was to become a significant factor in the provision of new housing, the saving in land would not be as large as might be expected. The towns of the future, if they are to become acceptable places in which to live, will have to provide more land for recreation, leisure activities, and amenity than was available in the cramped and overcrowded cities of the past. But, at least, for every family rehoused in regenerated city centres, there will be that much less pressure on land on the periphery, or in a new town created on a green field site elsewhere.

The supply of land for in-filling is beginning to dry up in the smaller towns and villages, so it does seem as if much of the land required for future housing will have to come from green field sites. In that case, should these sites be on the edges of large towns and cities, as a result, perhaps, of some easing of Green Belt restrictions; or will it have to come either from land on the fringes of existing medium-sized towns and villages, or from completely new towns? It seems unlikely that it will come from new towns on the scale of, say, Milton Keynes or

Bracknell. However, a few new developments on a smaller scale can be anticipated, though not in the already overpopulated South East.

Politically, any government would find it very difficult to sanction large breaches in the Green Belt policies of the larger cities, though there are undoubtedly some sites within these zones which were industrialised before the zones were defined, which are now obsolete. These could, and should, be developed for housing. In fact, there could be considerable merit in developing such sites, since local services will generally be readily available, and a completely new infrastructure is then unnecessary.

From the agricultural point of view, if green sites have to be given up, there is much to be said for taking out of production those already suffering disturbance from the proximity of a heavily built-up area, even though they might be situated in a Green Belt. This is especially the case if such sites are on poor quality land in the first place. Most of the supporting services for an increased population should then already be present, and a modest expansion should not require many new industrial premises, schools, hospitals and so on. This does not imply that there should be wholesale abandonment of Green Belt policies. Rather, there could be recognition that there may be a number of specific cases where a relaxation would have comparatively few disadvantages and undesirable effects, compared to a completely new development in the countryside, perhaps on the fringe of a smaller town.

As far as the general location of new building is concerned, it seems probable that it will come, as it has in the recent past, mainly from modest expansions of medium-sized towns, and even satellite villages within commuting range of such towns, rather than from the expansion of major cities. The remarkable change in the structure of industry over the past thirty years, from very large industrial plants to separate smaller component manufacturing units and assembly points, has enabled the increased population to be spread far more evenly over the country than was the case in the Industrial Revolution. Modern communications have, of course, greatly facilitated this change, though higher oil prices and the need

Intensive sheep farming alongside industry. This farm overlooks ICI at Billingham.

to control carbon dioxide emissions may make it necessary to restrict travel and transport in the future.

This trend seems certain to continue, with land being given up for small industrial estates adjacent to the developing towns. This will obviously mean the use of green field sites, both for industrial units and for housing the labour employed, once the original town has been in-filled to its maximum capacity. Again, from an agricultural point of view, it would be desirable for these sites to be located on land which is not of high agricultural potential. This might not, at first sight, seem to be very important if there is a surplus of agricultural land in any case. But, even in a surplus situation, the most efficient use of agricultural resources will be essential if the financial assistance given to the farming industry is to be kept at acceptable levels. This would not be possible if the best quality land was taken for development and farmers were left to farm the low grade soils.

The next question that has to be faced is where, geographically, this expansion of industry and housing should be allowed to take place. In recent years, major expansion has been primarily in the South East to North West belt, already briefly referred to on page 4. The South East, in particular, retained its relative prosperity throughout the recession that hit the Midlands and the North in the 1970s, and pressure on land there has increased enormously. In addition to that area, there have been two other growth points of significance. One of these is the East coast, where proximity to EC countries across the North Sea has acted as a stimulus to development. The other is the M4 corridor to Wales, where road and rail communications are particularly good; here completely new industries in the electronic field have sprung up, and many commercial companies and government departments, fleeing the high cost of London offices, have relocated their administrative bases.

Implications of the Channel Tunnel

To the already heavy pressure for land in the South East, has now to be added the imponderable effects of the projected opening of the Channel Tunnel in 1993. This is already accentuating demands for land in the South East for additional roads, rail links and new industrial sites with easy access to the Tunnel. It has to be accepted that very considerable areas of land will have to be given up, and that most of it will inevitably come from agriculture. Even existing houses and gardens will have to be sacrificed in the cause of easier access to Europe. It is not now just a case of losing good agricultural land under permanent urban development, it is one of losing some of the most attractive countryside in Britain in what used, in happier times, to be called the Garden of England. The tasks facing planning authorities here are almost insuperable, and there is certain to be both loss of amenity and of valuable land. Increased activity in areas bordering the London fringe could certainly jeopardise the Green Belt.

But the Tunnel may not only affect Kent. If the direct route across, or mainly under, central London materialises, it is certain to stimulate further development north of the capital, and up through the Midlands, exacerbating the problems which already exist around the major conurbations. There is likely to be most pressure in the vicinity of the motorways which will have direct road access to the Tunnel.

Competition will almost certainly force up the price of land, and this will have a knock-on effect on house prices, office rents and general overheads. High rents have already forced a significant movement of companies and offices out of London. Any further escalation could precipitate more departures, so increasing the demand for land and housing in those areas which are at present relatively undeveloped, such as the North. It is significant that Japanese car makers, seeking production bases in Britain, have often settled for such areas, where there is a supply of labour, and land and housing is relatively cheap. Further large-scale factory development is not very likely in the South East, because the price of houses is too high for workers to afford to live there. It is more likely to be located up the East coast, in the North, and in Wales.

But this does not mean that there will be much slackening of demand in the South East itself; the opening of the Tunnel will probably stimulate demand from Continental investors, anxious to establish themselves in Britain, for offices and light industrial development fairly close to the Channel.

So as fast as existing companies move out into less expensive locations, new ones could well move in to take their place, bringing new capital with them, and prepared to meet higher costs as part of the price to be paid for a presence in an influential area of Britain.

It would be a lot easier to estimate how much land is likely to be required for different purposes over the next ten years, if it was possible to forecast accurately the level of economic activity. This is never easy at the best of times. How often, for example, have even short-term Treasury estimates been proved faulty? The free market in 1992 and the Tunnel make it doubly difficult. Possibly two estimates should be made: one on the basis of a relatively buoyant economy, and one making an assumption that the opening of the British market to the full flood of competition from the European Community might lead to a recession. The following estimates are postulated. They cover the range between a reasonably thriving economy and a depressed economy, under four main headings of ways in which land lost from agriculture is likely to be used. The totals arrived at should provide a pointer to the possible effect on the agricultural industry, and to the intensity with which the land might need to be farmed. Alternatively, they might show that a lot more land will need to be lost from the agricultural sector, or taken out of production if large surpluses are to be avoided, whatever the state of the economy.

Domestic Housing and Industrial Building

Based on previous figures, an estimate of some 8,000-10,500 hectares is likely to be required annually for domestic housing, and a net figure of 3,000 hectares for industrial building. The lower total is more probable if economic conditions are stagnant. The industry figure would discount land which is redeveloped from obsolete industrial sites. Most of this 11,000-13,500 hectares would be coming out of agricultural use, though a small proportion would be derived from in-filling sites, such as gardens, paddocks and land not in mainstream agriculture, which does not feature in official returns.

Transport

Transport is unlikely to make the same demands on land in the next twenty-five years as it has in the past quarter century. There are a number of motorway projects to be completed, such as the M1/A1 link and additional roads from the Channel Tunnel. There will be more dualling of roads and by-pass construction. But road building is relatively slow, and the actual consumption of land per year is not as large as it might seem. The older motorways are now requiring very large sums for maintenance, which is certain to restrict, to some extent, the amount of capital available for new construction.

In addition to the road network, there is a serious need for more airport facilities. Whether this will come from the extension of existing major airports, or the construction of a completely new one, is a matter of conjecture, but whatever happens, the land needed will derive mainly from agriculture. A new airport would actually consume more land than the extension of those already in existence.

An annual *average* figure of 1,500 hectares should be enough to cover transport demands, though the actual figure may be higher or lower in particular years depending, for example, on Channel Tunnel works, or an airport extension.

The Extractive Industries

Five main products can be listed under this heading: gravel, sand, clay, cement and roadstone. The rate of extraction of the first four is heavily dependent on economic conditions and the state of the building industry. During the recession from 1975-85, the demand for minerals fell quite markedly from figures recorded in the post-war years, but this was followed by a resurgence in 1985 and subsequent years, until high interest rates provided a dampener again in 1989. Most of the major companies have considerable reserves in hand, and the main restraint is often planning permission to exploit the reserves, since many of the sites are in areas of scenic beauty. This is especially the case with gravel extraction in river valleys and quarrying in hill areas.

In a number of cases, the land can be restored for agricultural use without much difficulty. But, because of agricultural surpluses, it is likely that, to an increasing extent, sites will be restored for leisure activities such as water parks and nature reserves. However, from the agricultural viewpoint, this can be an advantage, as it means that leisure pursuits can be syphoned off into such areas, and thus relieve some of the pressure on better quality farmland. The likely closure of more coal mines will also release land for redevelopment, which could be directed to leisure pursuits rather than to agriculture.

As some of these sites are restored to farming, it is rather difficult to arrive at a net figure for the annual loss of land from agricultural production. But if economic conditions are reasonably buoyant, a figure of 1,500 hectares a year might be a fair average over a period of years. If there is a depressed economy, the figure would probably be nearer 1,000 hectares.

Water

One of the major consequences of increased population and higher standards of living has been a very great increase in the demand for water. More land will undoubtedly have to be found for its storage in future. From the agricultural aspect, this should, ideally, be on low grade land in upland areas, but many such proposals in the past have foundered on amenity grounds. This is surprising, since a well planned reservoir in the hills can both enhance amenity and encourage wildlife, especially bird life, whilst also providing for recreation in various forms. The sale of water companies into private ownership is not likely to affect very greatly the amount of land needed to meet future requirements for storage, though it might slightly increase demand. A figure of 400-500 hectares annually should be adequate to meet the needs of the industry, and this would be expected to come from land which is presently used for agriculture, though some of it might be of quite poor quality and low productivity in upland areas.

Conclusions on land requirements

So, the total loss of land annually in these four main sectors should be in the order of 17,000 hectares if the national economy is reasonably buoyant, but possibly down to 15,000 if it is sluggish. Initially, in the period up to the opening of the Channel Tunnel and immediately after, the demand for land could be greater, taking it perhaps to 18,000-19,000 hectares. Once the surge is over, it would be likely to drop again, and a twenty year average, up to say the year 2010, would probably not exceed 17,000 hectares. In addition, there could be a loss of land to afforestation of 1,000 hectares of new plantings annually, as distinct from replanting of felled woodland. This figure could increase if the Government pursued a very active afforestation policy as one method of trying to reduce food surpluses.

Robin Best estimated that land lost from agriculture from 1945 to 1975 averaged 15,700 hectares in England and Wales, and 2,000 hectares in Scotland; this included the years of rapid housing expansion, airport and motorway building and increased afforestation. So an annual figure of 17,000 hectares for England, Wales and Scotland should not be too wide of the mark. Best's estimate that by the year 2000, some 14 per cent of the land area of Britain could be urbanised, could well be something of an overestimate in view of the reductions which occurred from 1975 to 1985, and the downturn in the economy which started in 1989.

If the annual loss of land from agriculture for the next twenty years did average 17,000 hectares, this would represent a loss of only 0.15 per cent of land under crops and grass each year, which is not very significant in the context of Britain as a whole. It would mean that by 2010, approximately 15-16 per cent of the land area would be in the category of urban land. If that were so, the commonly expressed fear that Britain could eventually become a built-up area with little rural land left appears to be quite unjustified. Even though some regions such as the South East will inevitably become more highly urbanised, it is safe to assume that very large tracts of the countryside will remain completely rural in character, especially in the North and West.

The degree of urbanisation in some areas is the inescapable price that has to be paid for trying to accommodate, in reasonable conditions, a population of fifty-seven million people in a country the size of Britain, which has such a large area of mountain and hill land unsuitable for housing and industry.

What *is* important is that those areas which must be developed further for urban expansion and the better housing of the people, should be properly planned; that provision for recreation should be made; that good quality land for agriculture should be preserved; and that the scenic beauty of the undeveloped countryside should be retained and, if possible, enhanced, but not fossilised. It will require better national planning than in the past, if these objectives are to be achieved.

FOOD PRODUCTION AND THE EUROPEAN COMMUNITY

The national situation as far as agriculture is concerned has altered dramatically since the end of the Second World War. This has been due partly to the effects of the tremendous technical advances made in the scientific fields of crop and animal production and in mechanisation, and partly to the obligations imposed by accession to the European Community in 1973. Instead of being a single, isolated food production unit, Britain has now become one of a group of twelve, obliged to accept an overall control mechanism for regulating production and price support systems, with all farm products going into the total Community pool.

Due to technical progress and a series of good harvests, stocks of most of the major farm products were approaching surplus levels by the early 1980s in the European Community as a whole, even though Britain was not self-sufficient in quite a number of products. Stockpiles of food were also accumulating in the United States, and the disposal of surpluses on the world market was becoming very difficult, especially from the EC because of relatively high costs of production. Exports to other countries outside the Community were only possible through the payment of export subsidies, since world prices were well below EC internal prices. Such subsidies become an excessive drain on the Community's budget. Even though still in deficit with most farm products except cereals, the overall position was such that Britain also technically had a surplus of milk, dairy products, beef and sugar. So she had to submit to the curbs being devised in Brussels to restrict production, the first of which was the imposition of milk quotas in 1984.

This came as a great shock to British farmers, for ever since the severe food shortages and drastic rationing of the Second World War, Government policy and agricultural philosophy had been to produce as much food as possible within reasonable economic limits. This principle had been the mainstay of the 1947 Agriculture Act, partly as an insurance against shortage in the event of another conflict, but also because of the saving of foreign currency in the balance of payments account which was possible if a high proportion of the nation's food was produced at home. In pursuit of these policies, successive Governments of both political parties provided a series of grants for land improvement and reclamation, drainage, and so on, quite apart from the support prices agreed each year with the Farmers' Unions at the Annual Review of prices. Farmers became somewhat lulled into a sense of false security by this support, and came to accept that land should not be wasted and that maximum production was in the national interest. Technical developments in production and relatively good prices enabled all but the weakest farmers to re-equip their farms and bring them into a good state of fertility. The sense of security also encouraged many of them to borrow heavily to improve and expand their farming, on the assumption that future Governments would continue to follow similar policies.

Accession to the Community in 1973 led to a complete change in the methods of support for farming, with the higher prices which followed accession helping to stimulate production still further in the 1970s. But by the start of the 1980s, it was fast becoming apparent that the Community as a whole was moving into an era of surplus production, especially of milk and its products, beef, wine and cereals, and that steps would soon be needed to put the brakes on production, not least because of the very high cost of storing the surpluses until they could be moved onto the world market. The large amount of money being paid to farmers to produce these surpluses also had to be taken into account. Unless action to reduce them could be taken, they would bankrupt the Community.

Early attempts to reduce milk production by paying farmers to get rid of cows were not successful, nor were many bribed to get out of milk production completely. Those that did go out were mostly farmers with very small herds, so there was little reduction in overall supplies coming onto the market. So milk quotas were imposed on all dairy farmers in March 1984, which initially created many problems. But over the next five years, farmers were able to modify production methods, often by using grass and grass products in place of more expensive concentrated foods, so that the financial effects were not too serious.

Following milk quotas, levies were imposed on cereal producers in an attempt to reduce the area grown, with more price emphasis being placed on 'break' crops that were not in surplus. In 1988, in a further move to reduce production of cereals the Set-aside scheme was introduced, somewhat similar to that in America (see page 115). Under this scheme, farmers are compensated by means of payments on areas of their farms which they agree not to crop for a stated period. This land is supposed to be kept in a tidy state to prevent the growth and seeding of weeds. There has only been a modest take-up of this scheme by farmers, and certainly not enough to reduce surpluses of cereals materially, in the Community as a whole.

However, by 1989 the series of poorer harvests, combined with the constraints which had already been imposed, did reduce surpluses of cereals to a level not much above that required as a strategic reserve against further bad harvests, or against future famine in Third World countries. In spite of this temporary alleviation, expert opinion suggests that it would only need one or two good production years in the Community for surpluses to build up to their 1986 levels, and that measures will be needed to restrict production into the foreseeable future. So, as far as Britain is concerned, it can be assumed that for some years to come there will be no strong pressure to retain land in agriculture for food production, if there is a strong claim on it for another purpose.

The average annual increase in production of some 2 per cent due to improved technical efficiency, assuming that this continues, should be enough to make up for the loss of the 16,000-17,000 hectares of land going out of agriculture each year. If that area was all wheat, for example (which, of course, it would not be), it could mean a shortfall of some 80,000 tonnes, which is only 0.7 per cent of total production. Or again, if the land was all average quality pasture, it could represent a loss of some 50 million litres of milk, a very small proportion of total milk output, and of no great significance in the light of surplus dairy production. Improved technology could make up the difference in both cases.

So it should be safe to assume that loss of land is no longer a real threat to the future of agriculture, if it can be taken for granted that European or global wars will not, at some point in time, isolate Britain and turn her once again into an island fortress. This may not be an entirely safe assumption, but it has always been one of the main arguments in favour of membership of the Community. With the Channel Tunnel in being, it is difficult to envisage a 1917 or 1943 situation with Britain completely cut off from imported supplies. But, equally, it is very important that a healthy agriculture is maintained at levels of production adequate to ensure that enough food is grown at home to keep the population properly fed, without recourse to excessive imports of food which could be better grown at home. The critical phrase here is 'that a healthy agriculture is maintained'.

With an overall trade deficit of over £20 billion in 1989/90, nearly a quarter of which is represented by food and drink, it must surely be sensible to produce the maximum quantity of food at home, provided that this food is produced efficiently. Increased imports would inevitably worsen the already excessive trade deficit of this country.

This emphasises the importance of ensuring that agricultural production is maintained at near surplus levels, and that it should remain sufficiently profitable for efficient producers to be encouraged to invest capital in the land so as to ensure the continuation of a buoyant and efficient industry. Without continual investment of capital, no industry, be it agriculture, car production or electronics, can keep itself efficient and up to date. This is also highly important in so far as the management of the countryside is concerned. If there is no profitability in farming, farmers will be unable to spend money on conservation, and the whole rural environment will suffer. Those with memories of the 1920s can vouch for the sad effects on the appearance of the countryside of periods of agricultural depression. A well managed rural environment will only be achieved if the farming industry as a whole has adequate funds to devote to its maintenance, and to its enhancement.

It should also be reiterated that in maintaining a productive and viable agricultural industry the better quality land should where possible be retained for agriculture. This is where the higher potential lies in the event of a sudden need to step up production, and which, under a properly regulated system, requires lower measures of support.

This then is the background to the changing situation in relation to land use and modern farming systems, and the radical changes in outlook necessitated by the impact of surpluses of many farm commodities. Even if the loss of land from agriculture is now of less importance, farming is still a vital component of the rural environment.

But even if this is accepted as a matter of principle, it does not mean that methods and practices, developed as a result of the technological revolution of the past forty years, are necessarily desirable or acceptable on *all* counts. There has been mounting public criticism of some farming practices, so it is important that they should be reviewed in an impartial light. The following sections will describe in greater detail the changes that have been taking place in systems of food production and farming practice, and any action which might be necessary to maintain production.

Modern Farming Systems

PERIODS OF CHANGE

Throughout history, the practice of farming has never been a static process and evolutionary development has continuously altered methods and systems of production. The stimulus to change has been due partly to the need to produce more food as local populations increased, and partly to man's innate desire to experiment and to produce more in order to secure a better livelihood for himself and his family. In Britain, the pace of change has tended to fluctuate ever since the first major revolution occurred as a result of the Roman Conquest, and the introduction of a simple rotational system and improved methods of cereal production.

There have been two further periods of very rapid change. The first, in the eighteenth century, followed the enclosures and the introduction of new root and leguminous crops. This made it possible for farmers to practice more efficient rotations to improve the fertility of their land, and thereby to increase the yield of their crops and the productivity of their livestock. The second period of change, brought about by mechanisation and dramatic scientific advances in genetics, nutrition and disease control in both crops and livestock, has stretched over the past fifty years, and is still taking place.

Of these changes, the most recent has probably been the most significant in its effects, on both the life of the farmer and the countryside, greater even than the changes brought about by the enclosures and the new rotations, themselves very dramatic in their time. Such periods of rapid change have usually been followed by periods of slower development or even stagnation, mainly due to economic depression or political instability.

This was the reason for the long time throughout the Middle Ages when farming practices did not alter materially for several hundred years. But the clock has not actually been put back for any significant period. It seems unlikely that this will happen in the years ahead, in view of the projected expansion in world population, though temporary surpluses and economic pressures may slow down the rate of change. In fact, there are already signs that this may be happening. But it seems improbable that the remarkable progress in methods of production of the past fifty years will be lost; in the light of probable developments in biotechnology and genetic engineering, there may be further acceleration in the years ahead.

However much some of the more extreme traditionalists may hanker after the weed infested fields, waterlogged pastures, overgrown hedges, rabbit-dominated downlands and bracken-covered uplands — so characteristic of the depression years in farming from 1880 to 1935 — these could not return to any significant extent, except possibly in a few areas where soil or climate render the land marginal for crops or grass. No country in the world can afford to waste its natural resources in this way, least of all Britain with its great natural advantages for agricultural production, since the developing countries, with their own burgeoning populations to feed, will have little scope for exporting food. The changes that have taken place are virtually irreversible: just as the open field system of the Middle Ages could never return to modern Britain, nor a British Empire return to its former position of providing unlimited quantities of cheap imported food.

If this is accepted, as it must be by all rational thinking people, it will be all the more necessary to ensure that Britain retains a strong agricultural industry, especially in the run-up to the freeing of markets in the European Community in 1992. But this does not mean that, in maintaining such an industry, the best features of the countryside should not be retained and improved, and the destruction of environmentally desirable features be prevented.

Essentially, the latter objective can be achieved in two ways: firstly, through better education in the principles of conservation and environmental control of those who are the custodians of the land; and secondly, if necessary, through legislation designed to control the activities of those not prepared or willing to exercise their custodianship in a socially responsible way. Legislation should be avoided if possible, as it is always difficult to implement in such fields, and can often have an effect contrary to that for which it was designed. Education and persuasion are far better methods, legislation being only a last resort in problem situations.

So what changes have taken place in farming systems in Britain, and what pressures have led farmers to adopt the methods and scientific techniques that have become so widely practised in recent years, some of which have given rise to concern on environmental grounds? These need to be examined in some depth, before valid conclusions can be drawn.

TYPES OF FARMING SYSTEMS IN BRITAIN

Considering its small size, Britain is extraordinarily variable in its climate, topography and soils. The farming systems are equally variable; to a considerable extent they must be dictated by local conditions, even though mechanisation and scientific advances in crop and animal nutrition have in recent years gone some way towards reducing the restrictions imposed by climate and soil.

In very basic terms, there are really only three types of farming system, though there are an infinite number of variations within each type. These are, firstly, all grass, livestock producing systems; secondly, specialised arable systems where the sole output is crops; and thirdly, mixed farming systems, with a varying proportion of grassland and arable where the income is derived from both animals and crops. Traditionally, over many areas of Britain, the commonest system has been one of mixed farming. Exceptions include the upland areas of the country; heavy land farms in the North and West, which have always specialised in the production of milk, beef and sheep from grass; and in contrast, a relatively small number of farms in the Eastern Counties which have been, and still are, purely crop producing farms with no livestock on them at all.

The mixed farming category is itself extremely diversified, ranging from the large arable farm with only a small area of grassland and perhaps a beef or sheep unit, to the pre-dominantly grassland farm growing only a small area of crops, usually for consumption at home by the livestock — either a dairy herd, a beef breeding or fattening unit or a sheep flock. On the mixed farm, the plough may be taken right round the farm, alternating crops and grass leys (temporary grass swards of one to four years duration), or there may be more or less fixed areas of arable and grass because, for several reasons, it may not be convenient to plough a field, or to have it under grass. The first system is often known as ley farming. It has the dual advantage that the cereals and root crops benefit from the fertility created by the manurial residues left behind by the grass in the ploughed-in sward, while the livestock benefit from the vigorous growth of new grass in the reseeded ley, and from its freedom from animal pests and diseases.

But on many farms, the layout and topography does not permit this frequent rotation and makes fixed areas of grassland and fixed areas of arable land inevitable. In such cases, it is more difficult to maintain fertility on the arable block, though the advent of fertilisers and better cultivation techniques make this less of a problem than it used to be.

THE FORCES OF CHANGE

Without exception, economic forces have been imposing change to a varying extent in the structure and management of all types of farm. In general, this has led to greater degrees of specialisation and to a corresponding increase in the size of unit, whether the unit is a farm, a field or a livestock enterprise. Three main economic forces have influenced these changes: firstly, inflation with its consequent very large increase in the cost of all farm inputs such as labour, machinery, foodstuffs, fertilisers, chemicals, and of course, rent where a farm is tenanted. Even if the farm is not rented but owner-occupied, there may well be a large mortgage (due to the tremendous increase in the price of farmland), and interest payments can put great pressure on farm profits.

Secondly the level of prices received for farm products has not, in most cases, kept pace with inflation, thus imposing cost/price squeezes on the farmer. This was quite a significant influence under the old price control system which operated successfully from 1948 to 1973, but which terminated on entry into the European Community. Accession in 1973 brought some temporary relief, but the cost/price squeeze factor has become particularly relevant again as a result of surpluses in the Community from the mid-1980s onwards. Attempts to reduce production by the imposition of co-responsibility levies, and lower guaranteed prices for a range of commodities, combined with a lower price due to excessive stocks on the market, have all led to lowered farm incomes. But in Britain, inflation has continued to rise faster than in other Community countries. As a consequence, the squeeze on the British farmer has been particularly acute, and profit margins have continued to fall. Until the late 1980s, the EC support system seemed to favour cereal rather than livestock farmers, who felt the pressures most. But then, over-production of cereals, and more serious efforts to reduce the deficit on the EC budget, led to a much tighter squeeze on the cereal sector, and it was arable farmers who began to feel the pinch more than livestock producers.

Thirdly, there has been much pressure on the land tenure system, particularly from taxation policies. This has led, since 1946, to a virtual breakdown of the traditional landlord and tenant system, and the emergence of owner-occupation as the principal form of land tenure in farming. The fall from sixty five per cent of the land held on a landlord/tenant system in 1939 to only thirty five per cent in 1990 has had far-reaching consequences for farming. It is certainly one of the reasons for the rapid rise in the indebtedness of the industry as a whole over the past fifteen years. As landowners have sold up, either because of taxation demands (perhaps following a death), or because the income from farmland is generally low compared with what can be obtained in the City, so tenants have often had to borrow money to buy their farms, if they wished to remain in occupation. They have thus saddled themselves with interest payments which could be greater than the rents they were previously paying. The breaking up of estates has led to greater opportunities for the amalgamation of farms into larger and more economic units.

The overall effect of these economic factors has been to reduce quite dramatically the number of farms, and to increase the individual size of those that are left. The result has been that the national average size of farm is now nearly three times what it was at the end of the Second World War in 1946. The effect has also been to increase the size of each individual farm enterprise so that, for example, dairy herds are now three times the size they were forty years ago. The same is true of pig breeding herds. But at the same time the employed labour force, at approximately 95,000 full-time male employees, is only one quarter of what it was in 1946, while productivity per man has increased to an astonishing extent. However, to compensate for this, a farmer's investment in machinery, buildings and equipment has also been enormous. Increasingly over the years this investment has been with borrowed money, so that interest payments are now a major item in many farmers' accounts.

This is in marked contrast to the situation forty years ago when farmers, compared with ordinary businessmen, were very small borrowers, and a farmer who was known to owe money to a bank was viewed with some suspicion. In fact, in that time, farming has become very similar in many respects to an ordinary commercial business, and is regarded as such by the majority of those who are engaged in it. This is a good thing from the aspect of efficiency in management. But for some strange reason it is not regarded favourably by some sections of the public who seem to think that farmers should not be entitled to make profits, but should be kept on the land, almost as museum pieces, to keep the countryside in the sort of condition they imagine it was in during the pre-war years. Few who hold these views were alive then. If they had been, they might not be so keen about rural stagnation and depression.

The picture then of the years since the end of the Second World War has been one of great and intense concentration and specialisation, though this has not, by any means, been evenly distributed over all types of farming. Some aspects have changed more dramatically than others. A more detailed analysis of these changes in the different types of farming system now follows.

GRASSLAND FARMING SYSTEMS

Grassland farming is particularly associated with western and northern parts of the country and with upland

Typical Welsh countryside with trees and bushes on the lower slopes leading up to hill grazing. An area of high rainfall dominated by grassland and livestock farming.

farms, though examples can be found in most other areas under special conditions of soil type or in wetter river valleys. Upland farms, where beef or sheep production have always been the predominant enterprises, have been the least affected by change. This is because the climate and topography impose such severe restrictions on what can, or cannot, be produced profitably from the land, even with the aid of modern technology. As a result, there has been less scope, either for changing the components of the system, or for intensifying them to the same degree as has been possible under kinder conditions on lowland farms.

But changes certainly have taken, and still are, taking place on grassland farms of all types. In fact, it is on some of these marginal and hill farms, where change tends to be more apparent to the public, that modifications in farming practice have given rise to expressions of disquiet. This is because much of the land in hill areas has either scenic or recreational value, and any alteration in the appearance of the landscape stands out and is more easily visible. A considerable number of such farms may lie within the confines of National Parks, where they are very much in the view of people who may wish, for either aesthetic or other reasons, to preserve them in a completely 'traditional' state. There are, of course, many restrictions on what farmers may do with the land if it is in a National Park, but it is still possible for them to carry out some forms of improvement, such as reseeding.

Structural Changes on Grassland Farms

Change on grassland farms is really of two kinds: structural, and what, for want of a better word, could be termed technological. Structurally, upland farms have always been characterised by the smallness of the unit in relation to the quality of the land and the harshness of the climate, even though in the true hill areas the smallness of the farm may be offset to some extent by the rights to common hill grazing on moorland or mountain. The great majority are family farms, and have to support more than one working member of the family. But with the rise in the cost of inputs, potential output, even with a high standard of management, has often been insufficient, because of poor soil and climatic conditions, to provide an adequate return for the farmer, let alone for two or three other members of his family. So in upland and mountain areas, state assistance for both beef and sheep farmers has become an established part of the industry. This assistance has been directed as much towards maintaining the fabric of the countryside and preventing rural depopulation as it has been to supporting farmers in a difficult economic situation.

Of course, Britain is not alone in this. It is also the practice in other upland areas of the European Community, where large sums of money are allocated to what are known as Less Favoured Areas in order to retain land in production and keep the countryside populated. In effect, these payments are really equivalent to social subsidies designed to keep farming families on the land, rather than to increase production (which is unnecessary today in view of what is now obtainable from lowland farms). They are particularly needed in the Alpine districts of France and Germany.

But subsidies still do not prevent the smaller and more difficult units from being absorbed, as they become vacant either through death, retirement or business failure, into neighbouring farms in order to make them a more economic size for present day conditions. Farms may also come onto the market when old established estates are broken up on the death of a landlord, and further consolidation takes place in this way. Even in the upland areas therefore, farms are getting bigger, though probably not at the same rate as in lowland districts.

On lowland grassland farms the pressures for structural change are even greater, since they are not eligible for financial support. Here the need for additional land in order to spread ever-increasing overhead costs is inexorable. This applies especially to beef and sheep farms where the opportunities for intensifying production by technological means are rather less than they are in the case of dairy farms. So, on these farms too, there has been a steady increase in size in order to secure enough output to keep the occupiers financially solvent when margins between costs and returns contract.

Technological Changes on Grassland Farms

The second major change affecting grass farming units has been the move towards greater intensification of production in an attempt to wrest higher outputs from what are often still rather small farms. This intensification has generally involved the use of heavier dressings of fertiliser (especially nitrogen, which is so essential for a leafy crop, such as grass), and the application of herbicides to kill out weeds in infested pastures which may include attempts to remove bracken in hill areas. It may also involve the removal of earth banks or stone walls in order to make very small parcels of land into larger and better-shaped fields. This is essential for more efficient use of the larger machines now used for the conservation of grass, either as hay or silage, for winter feed. A further contributing factor is the very high cost today of maintaining walls and banks in a stockproof condition. It is labour demanding and the labour is often just not available to do the work.

Downland in southern England. Fields are large, but well-spaced-out copses and hedges provide for abundant wildlife.

So the farmer can hardly be blamed for removing some of them if it can make his business more viable, especially as some stone walls were only put there in the first place to dispose of the stones which were dug out of the land when it was first cleared. In the case of earth and stone banks in the West, a quite valuable increase in the area available for grazing can be obtained. But no sensible farmer in these areas will remove too many field divisions, as he values the shelter they provide for his stock in what are usually very exposed situations. However, it *is* important to be able to use machinery efficiently and speedily in such areas when weather conditions are fit, if good quality winter feed is to be obtained. This is all the more important because there will now be more stock to feed over the winter months, and high food conversion efficiency into meat or milk is vital if adequate profits are to be made.

In some marginal and hill areas, changes have gone further than just the increased use of fertilisers, weed sprays and a trend towards bigger fields. They include the reclamation of land, much of which was enclosed in the eighteenth and nineteenth centuries, but which reverted to scrub, gorse, bracken, heather or virtually useless grasses over the next hundred years, when prices for cattle and sheep were very depressed. When such land is reclaimed, the colour of the landscape changes from varying shades of brown, yellow or purple to the brighter, more uniform green of the fertilised grass field. In place of old rough grazings carrying just a few suckler cows or a small flock of ewes with an average of less than one lamb apiece, there are quite large ewe flocks or herds of beef breeding cows grazing within well-defined boundaries on green fields. This has allowed the farmer to make a better living, once the cost of the improvement has been paid off.

What concerns the conservationist, and to some extent the tourist also, who wishes to see the traditional patterns of brown and purple hills, and the variety of species of plant that the regressed land contains, is that such changes in the past have usually been accomplished, in part, with taxpayers' money in the form of improvement grants. What made matters worse was that the farmer then qualified for considerably larger headage payments on his livestock under hill assistance schemes, because of the increased stock-carrying capacity of his land after the improvement had been effected. Improvement grants on hill land are no longer available, so that situation is now a thing of the past. It is still possible for farmers to improve land if they are prepared to pay the whole cost out of their own resources, but with tighter margins, the returns are unlikely to justify the initial costs.

The dilemma has always been that the farmer, under prevailing economic conditions, really does need the extra output if he is to maintain even a very modest standard of living. Furthermore, the quality of the land is such that it has always been difficult to get an adequate return if he has had to provide all the finance himself. This is why grants were available, not only in Britain, but in all the countries of the Community, for this type of land improvement. But in the present surplus situation, there is certainly no justification for increasing production in this way, and assistance to hill farmers is given in other ways.

However, the arguments about hill land improvement remain as it *can* still be done. The farmer, who usually wants to improve only a very small proportion of his land, claims with some justification, that he is only affecting a very small area, and that there are still hundreds of thousands of acres of quite unimprovable hill land which the public should be able to enjoy, if that is what they want. Why, therefore, should he be debarred from bringing back into production a relatively few acres, of which he is the owner, or on which he is paying rent?

The conservationist or rambler could then reply, also with some justification, that in a few areas, such as Exmoor, there *are* no longer any very large blocks of unimproved hill land, and that if no restrictions are applied, the time may not be too far away when there will be none left for him to enjoy. Not only that, but the natural indigenous species of plants and animals will also have disappeared altogether. This is certainly true of a few areas, but in practice, in most of the hill districts there are still very large stretches of completely natural land. Unfortunately, much of this may be very isolated and almost inaccessible to the public, who remain quite unaware that it is there at all.

This is a question which will be discussed further in later chapters (see Recreation and the Countryside, and Conservation — Its Principles and Practice). Suffice it to say here that a considerable area in the uplands was improved in the twenty-five years following the end of the Second World War, but the costs of such work are now so high in the absence of grants that it seems improbable that it will ever be economic to do very much in the foreseeable future. What changes do occur will probably be more in the nature of using more fertiliser, accompanied by some alteration of boundaries, so as to facilitate better grazing control, and the use of machines for cutting grass for winter conservation.

Lowland Dairy Farms

On those grassland farms not situated in upland areas, intensification has proceeded rapidly, especially on dairy farms, and here a large input of capital expenditure has been required. The public today quite rightly demands milk of the highest hygienic quality. This has meant the pulling down of old, antiquated and insanitary buildings, and replacing them with modern milking parlours in which high standards of cleanliness can be maintained. The expenditure needed to achieve this has necessitated the adoption of high output systems if an adequate return on the invested capital is to be obtained. Many all-grass dairy farms in lowland areas are relatively small, so they have to rely quite heavily on purchased concentrate feedingstuffs if they are to achieve the output of milk required to provide a reasonable living. But such foods are costly to buy, so the fullest use of grassland has to be made if the feed is to be used economically.

Over the past twenty-five years, there has almost inevitably come a time on many smaller farms when modernisation of the buildings has become essential if the farm is to remain in milk production. The farmer then had to face two options — 'get on, or get out'. The fact that from 1960-1989 the number of milk producing farms declined from approximately 180,000 to 48,000 shows that a very large number of these farmers settled for the second option. Quite a number of the very small ones took advantage of the European Community's so called 'golden handshake' introduced in the late 1970s in an attempt to reduce the surplus of milk and milk products which was already beginning to build up. In practice, that measure was not very successful in reducing overall milk supplies, since many of those that went out of production were very small indeed, so their departure made little difference to total production.

Of greater significance was that those who opted to stay in milk production and modernised their farms in order to do so, had to increase the size and intensity of their herds to pay for the improvements and, in so doing, increased their output. As a result, milk production actually increased rather than declined, to the point that EC quotas had to be introduced in the spring of 1984, in an attempt to bring what was becoming a completely unmanageable situation under control. But the fact that the average size of the individual UK dairy herd increased from thirty-seven cows in 1960 to sixty-one cows in 1989, and the average yield per cow went up from 3,500 litres per lactation to 4,927 litres over the same period, shows that those farmers who did opt to stay in did, in fact, 'get on' in a highly efficient manner. The figures also show that attempts virtually to 'bribe' farmers not to produce a particular commodity may not have a very significant effect. This is because those who decide to remain in production of that commodity are generally the more efficient producers, and they will generally try to increase production still further in an attempt to maintain their incomes.

The introduction of milk quotas in 1984 forced all dairy farmers into a complete reappraisal of their production methods, and particularly of their grassland management. For the first time, output was restricted to a definite quantity, and they could no longer meet a squeeze on profits by stepping up their levels of production, which had been the standard reaction ever since the end of the Second World War. They could also no longer afford to buy in land in the form of imported concentrated feeds. The only alternative left to them was to make more efficient use of their grassland, and try to produce as high a proportion of their quota as they could from summer grazing and high quality silage during the winter months. In a way, the introduction of quotas was something of a blessing in disguise for the dairy farmer, though it did not seem like it at the time, since it forced him into higher standards of grassland farming. Since then, quotas have become a marketable commodity, either through purchase or leasing. This is having the effect of encouraging the smaller and less well-equipped farmer to sell his quota and go out of milk production altogether, while the bigger and more successful producers buy or lease quotas in order to increase production and the scale of their operations. The rather better prospects for sheep production (see page 42) provided an alternative enterprise for those grassland farmers who gave up milk, enabling them to get off the treadmill of intensive production which is so characteristic of dairying today. However, sheep production itself is becoming steadily more intensive and surpluses of sheep meat appear probable before long.

Lowland Grassland Farming and the Environment
From the point of view of the physical environment, increasing the size and intensity of the dairy unit on the all-grass lowland farm has two main effects. Firstly, it generally, though not always, leads to some increase in the size of the fields. Secondly,

it leads to an extension in the range of the farm buildings for the extra livestock which are being kept, and for preserving the grass for winter feeding, mostly in the form of silo barns or outdoor clamps on concrete bases.

Field size becomes important as herd size increases for two reasons. First, because the treading effect of large numbers of cows on a small field area in wet weather, especially on clay soils on which many grassland dairy farms are situated, can cause serious damage to soil structure. This then leads to lowered grass productivity. The damage can be minimised, though not completely avoided, if there is a larger area over which the herd can spread out from the gateways. Second, because the grazing season only lasts for some six to seven months a year, it is necessary to cut large quantities of grass at peak growth in early summer, and to conserve it as silage, or occasionally as hay, for feeding over the six lean months when there is little to eat outside.

If this conserved grass is to have a high feeding value (which it must if economies in the feeding of expensive concentrates are to be made), it is imperative that it is cut at the optimum stage of growth, and put into the silage clamp as quickly as

High speed mowers, both front and rear mounted on the tractor, enable heavy crops of grass to be cut for silage very quickly.

possible. This necessitates the use of large machines, capable of cutting, picking up and chopping the grass rapidly. So the field has to be big enough to allow large machines to work efficiently. It is not only the machines themselves that require room to work, but also the collecting trailers which have to work with them and need turning areas and so on. Thus, a reasonable field size is vitally important in modern dairy farming. This accounts for the modifications to field boundaries which have been made in recent years.

In most cases, hedge removal and the elimination of banks has been less on dairy than on arable farms, as cows need some shelter in rough weather. No dairy farmer wants *very* large fields, as the twice-daily collection of cows for milking during the grazing season can be time consuming in very big fields, especially at about six o'clock on a foggy morning when cows can get lost and left behind. A further reason for relatively small fields is the efficient provision of water. On a hot summer day, a dairy cow needs some 90 litres. If she is asked to walk too far to get it, she may not bother and her milk yield will drop accordingly. This factor alone imposes some limit on the size of fields on a dairy farm, whatever the size of the herd.

As dairy herds have increased in size, not only have larger buildings been required to house the cows for the six months when little or no grass is available for them outside, but larger storage areas for forage together with facilities for handling the extra volume of manure have also been needed. A dairy cow produces an average of 45 litres, or 10 gallons, of semi-liquid effluent or slurry each day. In large herds of 200 cows or more, this creates very considerable handling problems since the land is usually too wet for daily spreading during the winter months. The slurry can be handled mechanically in its semi-liquid form, but its storage demands the provision of pits, lagoons or towers in which it can be kept until the land is dry enough for spreading in the spring. This all requires extra land around the buildings.

On some farms with smaller herds, it is still possible to use the more traditional method of housing the cows in straw-bedded yards. Here slurry is absorbed into the straw and converted in situ into farmyard manure, which is then carted out and spread on the land in the spring. But this is no real solution today for the grassland farm. Straw is not available as a by-product as it is on a mixed farm, and the cost of purchase and transport is now too great for it to be really economic, except possibly if it is obtainable locally, and can be transported by the farmer himself at slack times.

So, with most grassland farms, and especially with the larger herds, cubicle or kennel housing of the cows has been almost universally adopted over the past twenty years or so. With this system, each cow has her own small lying area or cubicle. These are arranged in long rows on each side of a central passage, the dimensions of the bed being such that the cow will drop her manure into the passage, from where it can be scraped mechanically either into a slatted channel, or into a slurry pit twice a day.

On larger farms, cows are now almost universally milked in so-called milking parlours, and the old type of cowshed or byre is obsolete. The advantage of the parlour is that cows are milked in rotation in a small number of standings. This provides a very compact working area for the herdsman, and for the equipment required for milking. It enables one man to look after as many as 150 cows or more without undue stress, as the physical work is reduced to a minimum and little walking is required. But even though the cows are milked in a small building, there is still the need for concreted areas outside where the cows can be penned before milking, and for dispersal after milking. So the total space occupied by the cubicles, the milking equipment, forage storage and feeding areas, together with the storage of slurry is very considerable.

This has led to a situation where a range of modern dairy buildings begins to resemble an industrial complex set down in the middle of the countryside, rather than a traditional pattern of old farm buildings and barns with mellowed walls and red tiled roofs. This is undoubtedly a loss to the scenic value of the area, but there is really no way in which it can be avoided completely as it would be impossible on economic grounds to build today in eighteenth or nineteenth century materials. What *is* possible is to screen the buildings by the suitable planting of trees and to avoid high elevations on more exposed sites. Fortunately, even concrete and asbestos-roofed buildings can mellow surprisingly quickly and blend into a well-planted environment quite satisfactorily after a short space of time.

The other environmental problem arising with the large modern dairy farm is that of the disposal of slurry. To some environmentalists, this is of even greater concern than the changes in the size of fields or type of buildings. Even a modest-sized 100-cow herd, over a 200 day winter, will produce 910,000 litres of slurry, and many herds are twice or even three times as big as this. When this very large volume of material is spread on the land from the lagoon or tanks in which it has been stored, there may be some risk of surplus run-off into ditches or water courses if the spreading is followed by heavy rain. In addition, there can be a very obnoxious smell. Care must always be taken when spreading to avoid excessive application, especially on fields or on slopes adjacent to water courses. Also the spreader operator must always be careful not to spray it directly into a watercourse by mistake. One of the worst things that can happen is for an above-ground store to burst open, and pour the whole of its contents into the nearest watercourse, or even an adjacent village or town. This has occurred on more than one occasion. It can have a disastrous effect on fish and marine life, quite apart from impairing relationships between the farmer and his neighbouring community. Very heavy fines can be imposed for grave pollution of this kind, so from all points of view, it is important that the greatest care is taken in handling slurry.

There is only one sure way in which the smell problem can be solved, and that is to separate the liquid from the more solid material before it goes into store. But separators are expensive in themselves, while the equipment involved in collecting pits, channels, and pumps can also be costly. So only a small proportion of farmers can afford to install such a system. The reduction in the profitability of dairy farming, due to the restrictions imposed by quotas, makes it unlikely that separation will become a widely-adopted method in the immediate future, though it is possible that it may be linked to the production of methane gas plants if these become easier to operate. It is likely that modern technology will eventually provide farmers with methods of handling this material in such a way as to cause less contamination of the environment. Until then farmers will have to try to ensure that they only spread, if they live close to populated areas, when the wind is in the right direction, or is negligible in strength. Britain is not alone in this problem as any car traveller in Europe will appreciate. For some reason, the stench from the dairy farms in Germany and Holland when they are manure spreading always seems to be far worse than it is in this country. But that, of course, is no consolation to a resident in an English village when a local farmer decides to empty his pit, and the wind is in the wrong direction.

Grassland Beef Farms

For one reason or another, many grassland farms are not suitable for dairying. Such farms will carry beef breeding herds, beef fattening units, or sheep flocks, and very often a combination of at least two of these enterprises. Beef and sheep farms are not so heavily capitalised as dairy farms and are generally managed on a moderate input/moderate output system. The breeding cow can only produce one calf a year (until such time as reliable controlled twinning becomes possible through embryo transplantation), however well she is managed. So the only way at present of economically intensifying output on a beef breeding farm is to keep more cows on a given area by increasing grass production. Scope for this may be limited on the poorer quality land on which many of these herds are kept, and it is risky to use large amounts of fertiliser in attempts to increase production when the inherent capacity of the land is low, and profit margins are slender in any case. If profit margins disappear through a fall in the price of beef, then intensification merely compounds the loss. So under these conditions, farmers are seldom willing to risk using high input systems.

In the other main system of beef production where calves are bought in, grown on, and then sold as finished cattle ready for market over a fifteen to twenty-four month cycle, there is considerably more scope for boosting output. It is on this type of undertaking that intensification may begin to approach that found on the dairy farm. Such systems are often practised on mixed farms, as home grown barley and straw can then be utilised for feeding over the six winter months, but they can also be quite successful on the grass farm using silage.

The use of fertilisers is in that case often quite high, though seldom as high as on the dairy farm. Silage now plays an important role in winter feeding, so generally some modification of field boundaries will have to be made to facilitate its making. In general, however, it would be true to say that the intensification of beef production has not been responsible for creating any serious problems in regard to the environment.

Sheep Systems

The same would be true for sheep production on grass farms, even though intensification has been taking place, not only through increasing the number of lambs produced per ewe, but also by keeping more ewes to the hectare. These objectives are achieved through better management of the grass by using larger amounts of fertiliser, by stricter grazing control, and through the use of more prolific ewes, often crossbreds from hardy upland breeds.

Sheep used to be the poor relations of the farming world, left to fend for themselves for most of the farming year, and playing second fiddle to a beef or dairy unit. This was largely due to low prices for both meat and wool. The introduction of the so-called Sheep Meat Regime by the European Community, and the better prices for meat which followed, restored confidence, and in the past few years, some very large, specialised flocks of breeding ewes have appeared on both mixed and grassland farms. These are now receiving the high level of management input previously enjoyed by dairy and cereal enterprises. As a result, the output of sheepmeat from British farms has increased very significantly. Although Britain still imports a considerable amount of New Zealand lamb, controlled by EC quotas, sheepmeat is also exported to Europe and to some countries in the Middle East.

The change that has occurred in sheep management is very noticeable to anyone driving through the countryside in the early summer when very heavy concentrations of ewes and lambs can be seen on the pastures, with much higher rates of stocking than would have been seen even fifteen years ago. Sheep have one great advantage over cattle in that they graze very close to the ground. This encourages the growth of white clover which has the capacity for fixing nitrogen in the root nodules, thus reducing the need for heavy applications of nitrogenous fertilisers to boost grass production. Indeed, these can be somewhat deleterious in sheep farming, as they can actually depress the white clover, on which sheep can thrive.

The impact of sheep intensification has not, perhaps, been quite so noticeable on lowland farms as it has been in the uplands where pasture improvement has a more dramatic visual effect. There is, incidentally, a very strong inter-dependence between upland and lowland sheep farmers. This is because: firstly, many lowland breeding ewes are the progeny of hill bred ewes mated with a lowland ram, and they inherit the qualities of toughness, hardiness and good foraging ability from their mothers. They grow well when transported to the kinder conditions on lowland farms, where their breeding capacity also increases. Secondly, hill-bred lambs cannot be finished adequately for the market in their local environment; they too are transported to lowland farms where they are finished in the autumn or winter months on grass or arable land, providing additional income for the lowland farmer. This is why it is very important for the balance of farming to maintain a healthy and viable upland sheep industry. It is also why farmers on upland and hill farms have been anxious to improve their grassland, so as to produce more lambs for sale. The upland farmer has a ready market for his culled ewes and lambs, while the lowland farmer gets healthy and prolific breeding stock, and store lambs to finish for the market.

All-Grass Farms and the Environment — A Summary

Generally speaking, the changes that have taken place in grassland farming have probably given rise to greater concern on upland than on lowland farms. This is especially so when reclamation has been involved, when the contrast in colour from brown to bright green is very striking. When considerable areas have been reclaimed, the whole appearance of the countryside can be altered, the previously wild and colourful scenery tending to take on an air of domesticity. This may be a good thing for those who are seeking to make a living from the land, but it does not suit those who have come for recreation or to study the local environment.

If the economic situation had not changed so dramatically due to overproduction, there might have been a need, either for legislation or for voluntary contracts, to restrict further reclamation in certain areas. Exmoor is a case in point. But with the withdrawal of grants, and the increasing cost of reclamation, it seems likely that extensive ploughing and reseeding of hill land will not be a viable proposition in the future. Indeed, if there is a depression in farming, brought about by overproduction, the reverse process could well take place quite rapidly, namely a reversion to rushes, sedges, heather and dwarf gorse. It is surprising how quickly land can revert, once dynamic management is withdrawn. There is usually a large store of ungerminated seed buried in the land when it is ploughed, and the plants growing from it very quickly re-establish, if they are not controlled. So the controversies of the past twenty-five years could diminish or disappear completely in the future.

On lowland grass farms, modern methods mean larger parcels of land, concentrations of bigger farm buildings of a less picturesque appearance than those erected in earlier centuries, and deeper coloured green fields carrying many more animals than in the past. They can also mean the possible contamination of waterways through the run-off or seepage of nitrogen from fields receiving larger dressings of either slurry or nitrogenous fertiliser. In addition, they can sometimes mean unpleasant smells from the spreading of the slurry, or very occasionally from the silage clamp itself if the fermentation has been at fault.

This may seem a formidable list of things which *can* affect the environment, but it must not be assumed that they will be associated with *all* grass farms. Of course, they will not. But they can be seen in varying degrees and very occasionally to the extent where they begin to cause public concern. It is in these cases that the reasons for what is happening need to be understood and considered in a dispassionate way without the intrusion of excessive emotion. The vast majority of farmers have no wish to offend their neighbours, or those who use the countryside for a wide variety of purposes, and they will do all they can to meet legitimate grievances. There are, of course, as in any walk of life, a small number of 'black sheep' who do not want to conform, and who have little or no concern for public opinion. For these, legislation is probably the only way to control obvious excesses, such as tree felling. For the remainder, it is important that they should take note of public disquiet, and attempt, not only to meet criticism where it exists in a spirit of co-operation rather than aggression, but also to pursue their farming systems in such a way as to cause the least interference with public enjoyment of the countryside. The public, for their part, must accept that the farmer has a right to make a decent living, and that some of his actions may be essential if this is to be achieved, while others may be determined by Government policy and regulations.

ARABLE SYSTEMS

At the opposite pole to the all-grass farm is the arable farm. Here, virtually all the land is under the plough, and livestock, if they are kept at all, are likely to be units of pigs or poultry, or perhaps of intensively managed beef cattle housed throughout the year. In the past, purely arable farms were comparatively rare since the traditional pattern was one of mixed farming with quite a high proportion of farm output in the form of animal products. Over the last twenty-five years or so, there has been a marked increase in the number of specialised arable farms, mainly due to the fact that cereal prices were consistently more attractive to the farmer than those for livestock products, with the possible exception of milk. So the natural reaction was to reduce the area of grassland and to grow cereals instead. A secondary, but very important, factor was the relatively low labour requirement of an extensive arable system based on cereals, when compared with livestock management. So, in areas where there was severe competition for labour, or where housing has been difficult, farmers have tended to switch over to systems which can be operated with a small input of labour. These were mainly in the Midlands and the South where the climate is more suited to arable crop production.

In modern farming, there are really two quite distinct types of arable farm. Firstly, there are those which grow a wide variety of crops in a fully rotational system, e.g. wheat, barley, potatoes, sugar beet, beans, peas, oilseed rape, and perhaps vegetable crops on a large scale for freezing, canning or the fresh market. This is the traditional type of arable farming developed over the past two hundred years. Concentrated mostly in the Eastern Counties, it can also be found in relatively small pockets in other parts of the country well favoured by soil and climate, for example East Lothian. Such systems have a high labour requirement, though this can often be met by the employment of casual labour for specific tasks. In these areas there is a tradition of gang labour, hired by farmers as and when it is needed. The permanent resident staff can in fact be quite small, having been drastically reduced on many farms in recent years as wages have increased.

Secondly, there is now the arable farm which relies almost exclusively on cereals in what is virtually a system of monoculture, though the recent introduction of oilseed rape has provided a new crop, which not only alternates very well with cereals, but which can also be handled by the same machinery. But even with the inclusion of oilseed rape, the system is still very highly specialised. One of the most significant changes in British farming since the end of the Second World War has been the increase in number of this type of farm, and the spread of similar systems into areas which used to be dominated by mixed farming. Both types of arable farm need to be discussed in more detail since the environmental problems associated with them are not the same.

The Traditional Arable Farm

The conventional Eastern Counties arable farm has not really altered very materially for a hundred years or more, even though new crops, such as sugar beet in the 1920s, vegetables for freezing in the late 1950s and oilseed rape in the 1970s, have widened the range of possible combinations. Farms of this type have always been farmed very intensively for high yields on quite strict rotational lines, even in the years of depression. They had to be since the value of the land and the rents demanded for it have always been high relative to other areas. Output from every acre under crop has had to be maximised, with no land wasted and hedgerows kept to a minimum. To a non-agricultural eye, the appearance of the countryside has always been, and still is, somewhat unattractive. This is especially so in the Fenlands, where the banks bordering the dykes and occasional scattered trees are the only features of the landscape, apart from the variety of the crops themselves. Wildlife tends to be concentrated on the dykes and their surrounds as there is little other land left uncultivated.

Both the intensity of production methods and the size of crop yields have increased very considerably over the last thirty years, due not only to genetic improvements in the varieties grown, but also to a much heavier use of fertilisers, insecticides, fungicides and herbicides. Land prices and rents have soared, and this has meant that there has been no real alternative but to increase production as far as economically possible. Of course, if the cost of inputs continued to rise, but prices remained constant or even declined, it might at some point become more economic to embark on a more moderate input/moderate output system, even though the high rents might make this difficult. But that point has not yet been reached for the types of crop grown on these farms, and it seems unlikely that it will be reached in the immediate future. High input/high output systems will almost certainly continue on this type of farm, unless restrictions on the use of nitrogenous fertiliser are imposed. In that case, some form of payment would be needed to help the farmer meet his rent bill or the interest payments on his land.

The main environmental concern arising from the cropping policies on farms of this type is not so much interference with wildlife or the appearance of the countryside (neither of these has ever been a strong feature), but residues from chemical applications, which are essential ingredients of the system. This particularly applies to nitrogen, the main component of the fertilisers used in the production of root and vegetable crops. These rely for their yields on a very large leaf area, for which high levels of available soil nitrogen are needed.

Since this type of production is generally located on light-textured land (for ease of working round the year, for harvesting of crops under wet conditions, and to allow for root expansion), the risk of leaching of surplus nitrogen into watercourses is increased. The result can be excessive growth of algae and weeds in the dykes and ditches, and a consequent increase in eutrophication, especially if phosphate levels in the water are also high. This affects the degree of biological activity in the water, quite apart from increasing the cost of maintenance of the dykes, which must be kept clean to ensure the removal of surplus drainage water in the winter. But far more important than this now is the actual level of nitrogen in the water which may exceed the standards laid down by Brussels for potable water. As many domestic water supplies are drawn from surface reservoirs and rivers, the higher nitrogen levels in the water draining into them is presenting a very serious problem both for the water companies and the Government, as it is very expensive to reduce it to the required level. It is generally recognised in Britain that the EC level is extremely strict, but as members of the Community, the regulations must be complied with.

It is difficult to see any immediate solution to the problem, which seems certain to deteriorate still further, since nitrates can take many years to percolate down to the deeper aquifers. Either the water authorities will have to install expensive extraction plants, or the Government will have to impose some restrictions on farmers on the quantity of fertiliser applied and the time of year at which they do it. Already farmers are being advised not to apply any fertiliser containing nitrogen in the late summer and autumn, because of the ease with which it can be washed down in the autumn rains. The majority are acceding to this suggestion and have adopted this practice. The Government has also defined certain 'Nitrate Sensitive Areas' (NSAs). Where farmers have been asked to reduce applications, and make other changes in their farming practices, tests will be carried out to assess the effect of these measures on the ground water. Farmers will obviously lose some yield in their crops if the level of application is much reduced, and compensation can then be paid for such a loss. Of course, farmers are not the only people who are implicated, since considerable quantities of nitrogen find their way into watercourses from sewage plants and from industrial effluent. It would therefore be desirable to tighten up on these sources of pollution as well.

The whole nitrogen question is still obscure, as can be gauged from the following extract from the 'Rothamsted Guide to Soil and Fertiliser Nitrogen'. 'The nitrate that leaks (leaches) from arable land comes mainly from the breakdown of soil organic matter and crop residues. *Very little comes directly from unused fertiliser applied at recommended rates.* It will take decades for any cut in fertiliser use to have an appreciable effect on nitrate leakage. Ploughing up old grassland causes a substantial release of nitrate over many years. This, rather than fertiliser use, might in some areas be the main reason for the increase in nitrate concentrations in water supplies from aquifers. Research at Rothamsted and elsewhere has established the basic principles for abating the nitrate problem. We are now in a position to apply these principles to a wider range of farming systems and to anticipate nitrate problems arising from changes in farming practice and land use. Much will depend on sound farm practice backed up by good professional advice. Hasty solutions that do not consider the full complexity of the problem will not be effective.'

It follows from the above that the establishment of Nitrate Sensitive Areas is not really likely to have much effect on the nitrate-in-water problem in the short term but, of course, it shows that the Government is 'doing something'; even any quite small results could be of value in helping scientists to build up a more complete picture of the issues involved.

Wind erosion is an environmental problem of an entirely different nature in some relatively small areas of the light arable land farming districts in the Eastern Counties, including the Fens. This occurs in dry weather, usually following spring cultivations for the production of fine seed beds prepared for the sowing of small seeds. If these cultivations are followed by

Wind eroded sand particles blocking a ditch in West Suffolk.

a spell of dry weather and strong winds, as can often occur in February and March in Eastern England, the surface dries out rapidly before there is adequate plant cover, and thousands of tons of top soil can be blown off the surface into the surrounding hedgerows, dykes or roadways. In really serious 'blows', smaller seedlings can be uprooted, or the tender leaves of the young plants ripped to pieces by the blowing sand particles.

A variety of methods are used to try to counteract this problem, so far as it is possible to do so. These may involve the planting of shelter belts, intercropping with taller plants which can be sprayed out at a later stage, or possibly leaving a cereal stubble from the previous crop on the surface and using minimal cultivations or direct drilling for the new crop. But no method is completely successful against really strong winds in a dry time. Blowing has always been a problem in these areas, and there is no evidence that it is getting any worse as a result of the more intensive farming methods now being used. In fact, the reverse may be the case, since with the increased cost of land, seed, and fertiliser, more efforts are now being made to try to avert the financial loss involved. This may be very serious, as valuable top soil is lost when the blow occurs, quite apart from damage to the crop if it has come through the ground, which may then require resowing.

On this type of farm, therefore, increased pressures for higher production have not materially affected the environmental situation, except in relation to the use of nitrogen, and possible residues from pesticides, especially insecticides. Pesticides are used extensively to protect high value crops such as sugar beet, potatoes and vegetables from insect and fungal damage, which not only affects yields, but can also seriously impair marketable quality and uniformity. This is especially important today when so many vegetable crops have to be grown to very strict specifications laid down by the supermarket chains, who are now the principal buyers of the produce. Any crop which falls even very slightly below the specification may be refused out of hand, and can be very difficult to sell, because there are now so few buyers left in the market. In the case of potatoes and sugar beet, control of insect pests is vital, since they will frequently be vectors of plant viruses. These can drastically reduce yields and quality if they become established on a wide scale across the crop.

There is no strong evidence of any serious water pollution from pesticides in Britain, though there are lessons to be learned from America, where contamination of drainage water has occurred as a result of excessive use. Significant accumulations in the soil are also considered unlikely, as many of those in common use are quite quickly broken down in the surface layers and because the same pesticide will not be used on the same field too often (see pages 86 and 93).

SPECIALISED CEREAL FARMS

Reference has already been made in connection with grassland farming to the increase in specialisation in farming enterprises since the last war. Nowhere has this been more marked than in cereal production. Cereal crops have always been grown widely on the calcareous soils of the South and East of England, in many heavy land areas of the Midlands, and in drier districts elsewhere. The cereals grown were usually wheat and barley, which require warmer and drier summers to reach maturity than oats, a crop found more extensively on mixed farms in the West and North, where it was grown primarily for livestock feeding. With the decline in the number of heavy working horses, and the introduction of higher yielding varieties of both wheat and barley, the production of oats declined quite dramatically. Not in surplus, it is now being grown again to some extent as a break crop from wheat and barley. There is also an increasing demand for oats, not only from the ever-growing number of horses kept for recreation, but also for use in the health food market.

Barley largely took over land previously devoted to oats, while wheat spread out into districts where it had never been grown before. Modern varieties of these two crops have much stiffer straw than the older types, and this enables them to be grown in higher rainfall districts without so much risk of lodging (being flattened) before harvest, which causes a serious loss in yield. In addition, the modern combine harvester with effective pick-up reels has largely removed the difficulty and worry of dealing with laid crops. The other advantage of a stiff straw, of course, has been the ability to use much larger applications of nitrogenous fertiliser without so much risk of lodging. This allows the farmer to exploit the genetic capabilities of the crop to the fullest extent.

So farmers specialising in cereals can now be found in some areas which were previously dominated by grass and livestock farming, while the proportion of cereals on mixed farms is often far higher than it used it be. Indeed in some cases, rotational farming has disappeared altogether, to be replaced by continuous cereal cropping, or perhaps by cereals broken occasionally by oilseed rape or a leguminous crop, such as peas or beans.

Contrary to general opinion, such systems are not really new. The late Arthur Baylis grew cereals only, with an occasional full fallow, for some forty years around the turn of the twentieth century in Berkshire, while Philip Chamberlain grew barley continuously in Oxfordshire between the wars. Quite a number of farmers were experimenting with continuous cereals in the late 1930s when tractors began to replace horses, and larger implements and combine harvesters were coming onto farms. The advantage was that both tillage and harvesting operations were greatly speeded up, and labour costs reduced quite dramatically.

This was at a time when cereal prices were very low and land was cheap, and the returns from beef and sheep which might have competed with cereals were at rock bottom. The main attraction of cereals at the time was that the larger farms could be managed with a very small labour force and the costly handwork involved in livestock farming avoided. The replacement of men by machines helped the farmer to compete more effectively with the flood of cheap grain produced under prairie conditions in North America and Australia.

But, in practice, many of the attempts at continuous cereal growing before the Second World War ended in failure. This was either through an uncontrollable build-up of weeds, especially in continuous autumn sown wheat systems, or through fungal infections which drastically reduced the yields of both wheat and barley, to the extent that production became uneconomic in spite of the system's low costs. In those days, there were no effective herbicides and no systemic fungicides to help in controlling these problems. Under the conditions then prevailing, farmers had no alternative but to revert to more conventional rotations, and in many cases to introduce grass leys again between the cereal crops, even if this did mean very poor financial returns. But milk prices had by then stabilised, due to the establishment of the Milk Marketing Board in 1933, so this did provide a potentially profitable alternative on those farms which had suitable buildings for dairying. The inter-war years saw the introduction by Arthur Hosier of the portable outdoor milking bail, the precursor of the modern milking parlour, onto larger arable farms, in an attempt to find a more profitable way of using grass leys than grazing them with cattle and sheep.

From the early 1950s onwards, a number of quite independent technical developments combined to make intensive cereal growing systems more attractive again. The earliest of the systemic herbicides had been introduced during the war years, and after the war a whole range of new compounds rapidly made their appearance. These were capable of controlling the majority of weeds which had previously proved so troublesome to the earlier continuous growers, though a few, such as wild oats and black grass, still remained to cause problems. But by the 1970s even these were reasonably controllable with newer herbicides. Similar developments were also taking place with fungicides, and it soon became possible to control most of the diseases that had caused such severe losses in yield before the war.

Almost as important were the rapid advances in mechanisation, not only in the sheer size of the tractors, implements and harvesters, but also in their speed of work. This enabled field operations to be carried out very quickly when conditions on the land were suitable. It enabled fields to be prepared for the sowing of autumn crops in good time after harvest, so that new crops got a flying start, and were better able to stand up to hard winter conditions than if they had been sown later. When the work had to be done with horses, timeliness was never possible over a whole farm since cultivations were so slow. All weed control had to be achieved by cultivations which were always at the mercy of the weather, especially on heavy land farms. As a result, planting of both autumn and spring sown cereals would always be late on some fields. The use of chemical weed control has made an immense difference to yield, both because of timeliness of planting, and also the early removal of weed competition during the crop's growing period.

The introduction of the large, wide boom sprayer with adjustable height has really revolutionised cereal growing, enabling the farmer to apply, when conditions are fit, a wide range of compounds for the control of weeds, diseases and pests and even the final height of his crop. Often two or more chemicals used for different purposes can be mixed together; this saves time and money in application, and too frequent passage over the field of heavy machines which risk damaging soil structure on unstable soils. The end result of all these developments is that it has now become possible to eliminate most of the factors previously standing in the way of growing cereals continuously, and farmers have not been slow to avail themselves of the new technology.

The plant breeder, too, has had a significant part to play. Not only has he provided new cereal varieties with much greater yield potential, but he has also given these varieties much stronger and stiffer straw, and a certain amount of built-in genetic resistance to some diseases, such as mildew. This helps to reduce reliance on purely chemical methods of control, and so reduces the cost of production. The hope for the future is that plant breeders, through developments in biotechnology and gene manipulation, will be able to build in much higher levels of genetic resistance to both pests and fungal diseases. If this can be achieved, many of today's chemical sprays could become redundant. There would then be far less concern either about any possible long term effects of these on the environment, or of plants developing resistance to them, something which is already a possibility.

Soil Structure in Continuous Cereal Production

Another most important factor in continuous or intensive cereal production is the maintenance of good soil condition, which is largely dependent on maintaining a satisfactory level of organic matter. So long as the potential yield of a cereal variety was only 2.5 to 3 tonnes per hectare (1 tonne to the acre), there was a serious risk of a loss of stucture on many types of soil due to a failure to return sufficient organic matter to the land through crop residues. This was especially critical on lighter soils and on some silts and heavy clays. If this happened, there was a vicious downward spiral of deterioration. Poorer root systems and stunted crops caused by bad soil structure led to smaller amounts of organic matter being left behind. This, in turn, led to a worse condition the following year with even poorer root systems and so on, until yields became quite uneconomic.

One of the main benefits derived from the introduction of wheat and barley varieties with a yield potential up to three times that of the older varieties, has been the capacity to grow much larger root systems; in fact, present yields could not be obtained without them. These appear to return enough organic matter to the soil to maintain structure on many soil types, and to permit long runs of cereals to be grown without undue deterioration in this structure. This applies especially to autumn sown wheat and barley. It is not so true of spring sown cereals on lighter and drier soils as these crops are shallower rooting, and in dry seasons only produce relatively small root systems during their restricted growth cycles. These may be inadequate to maintain a safe level of organic matter. Organic matter is also, of course, broken down more quickly by oxidation in lighter soils than it is on heavier land, so that levels drop more quickly.

The higher yield potential in cereals, however, is only realised through the use of considerably higher levels of fertiliser, especially nitrogen. This in itself has a secondary beneficial effect on soil fertility. In addition to stimulating plant growth in the current crop, higher soil nitrogen levels can lead to a more rapid breakdown of organic matter and crop residues, thus ensuring a quicker recycling of nutrients and a more biologically active soil. The widely publicised statement that the application of high levels of fertiliser in some way reduces biological activity in the soil, thus 'poisoning' it, does not appear to be valid in practice. If it *was* true, current crop yields, particularly in continuous cereal systems, would be impossible to achieve.

New Techniques and Cereal Growing
So the introduction of such a large number of new techniques over the past twenty-five to thirty years has made it far more possible to farm intensively with cereals over quite a wide range of soil and climatic conditions. While it may not be possible to obtain the really exceptionally high yields which can be achieved under rather more conventional rotational systems, quite satisfactory yields are frequently possible, often with a lower level of fixed (overhead) costs, as the investment in labour and machinery is generally lower than in more complex cropping systems. Concentrating on only one type of crop enables the farm to be operated with a narrow range of machinery, even though the individual machines may be very large. They allow the work to be done as quickly as possible by a very small labour force, and when conditions are fit.

To give an example of the yields that are possible: on the farm recently managed by one of the authors, one field averaged 5.3 tonnes of wheat per hectare (42 cwts per acre) for a run of twenty-one years of continuous cropping. This is certainly not a particularly good yield by today's standards, but the run started in the early 1960s when yields were much lower in any case, and in the early years they suffered from weed competition, since the appropriate range of herbicides and higher yielding varieties were not available. The interesting point to note is that for the last five years of the run (which were not in any way outstanding for wheat yields), the average yield was 6.5 tonnes per hectare (52 cwts per acre). This must surely disprove the assumption that yields are bound to deteriorate over long runs of cereals, and that soils will become 'farmed out' under such conditions. The above example is by no means unique, results from research farms providing similar evidence. The idea that soils are ruined by such systems are based on pre-war and immediate post-war experience. They simply do not hold water under present-day conditions.

But in spite of savings in some costs in such a system, intensive cereal growing is no longer a low cost enterprise, nor is it a lazy system as some of its critics would suggest. If it is to be successful, it requires a high degree of technical knowledge and application. It also demands a considerable input of fertiliser, fungicide and herbicide and possibly a straw shortener as well. But even so, the combination of technical aids and powerful machinery for cultivating the land as soon as conditions are fit after harvest, plus the high yielding varieties of wheat now available, made this an attractive arable system for a farmer, if he had land which grew autumn sown cereals well and the climate was suitable for cereal crops but not for grass owing to the risk of summer drought.

There is, however, another factor which now has to be taken into account on this type of farm; namely, the ban on straw burning which becomes operative after the 1992 harvest. On many farms which have been practising systems of this kind, much of the straw has been burned in the field after harvest. This is because: firstly, there was no use for it on a farm without livestock; secondly, there would not be adequate labour to cope with it; and thirdly, burning had a very valuable hygienic effect through the destruction of diseases carried over from year to year on the straw and stubble. This was probably worth the cost of at least one fungicidal spray, and possibly more. In the future, the only alternative will be either to chop it up into small lengths behind the harvester and plough it in, which is difficult on very heavy textured soils, or, to find the extra labour to bale it in the hope that a market can be found for it. Most farmers will probably opt for the first alternative, unless new industrial uses can be found for straw. The advent of anti-burning legislation may cause a number of these farmers to give up the system. They may either go back to livestock farming with leys and a proportion of cereals, or put some of their land into Set-aside, or other alternatives such as tree planting. In either case, it could lead to quite a sizeable area going out of cereal production.

The second factor which made continuous cereal growing attractive up to the mid-1980s was the farm price structure, both under the old British system of acreage and deficiency payments and, even more so, under the Community system after 1973. This certainly favoured cereals rather than livestock products, with the notable exception of milk. The cost of modifying and updating buildings for livestock rose to a high level from the late 1960s onwards, and this, coupled with the higher labour demand for livestock, gave the farmer very little incentive to expand beef and sheep enterprises. In fact, all the pressures were towards giving up livestock altogether if the farm was suitable for cereal production. He could then dispense with high overtime payments for weekend work with the livestock, and was spared the constant need for vigilance and supervision which livestock farming entails.

When prices were buoyant through the 1960s and 1970s, intensive cereal production did have much to commend it from the aspect of both profitability and efficient land management over large areas of the South, the East and the Midlands, and even in some areas of the North where the rainfall was not too excessive. It is not easy to blame a farmer for adopting a production system which the prevailing price structure makes attractive. In the last resort, or even in the first resort, he does have to be a good businessman if he is to survive and get an acceptable return on invested capital.

But from the mid-1980s the situation began to change rapidly due to the surpluses of wheat and barley, which began to accumulate in the EC after a series of good harvests, and as a result of the rapid improvements in technology. This led to drastic attempts by the Community to curb production. This had a special importance because of the effect of the large stockpiles of American grain which were keeping world prices low, and making it necessary for Brussels to spend huge sums of money in subsidising exports in an attempt to dispose of its own surplus. Guaranteed prices for cereals, purchased through the Intervention system, were reduced and co-responsibility levies were introduced which had to be paid by cereal farmers. In 1988, the Set-aside scheme was launched.

The overall effect of these measures was to reduce both confidence and profitability in cereal production. Quite suddenly the cereal boom had come to an end, even though the writing had been on the wall for some years, and it began to be difficult to make adequate profits out of grain. To compound problems for the farmer, the yield of cereal crops fell in 1987 and in 1988, largely due to bad weather both in Britain and on the Continent, and the harvest of 1989 was only about average (taking the Community as a whole). So by the beginning of 1990, the surpluses, which had caused such alarm some five years before, had virtually disappeared, and stocks were down to a figure not much in excess of that needed as a strategic reserve. By the spring of 1991, however, surpluses of cereals and dairy products in the EC were again back at high levels.

This illustrates only too well the problems that can face planners when trying to control production over an area the size of the European Community. For example, was this reduced production a short term or a longer term trend? If it were a long term trend, then the actual need for a Set-aside policy would come into question. It might then be more economic to pay money in subsidising a moderate level of exports than to pay large sums to farmers simply to take land out of production. Looking at the problem in 1991, it would seem to be only a short term trend, since one really good harvest in Europe virtually filled the surplus stores once more. This is because, in spite of the efforts to take land out of cereals and reduced profitability, the area still remains high, the reaction of farmers being to try to compensate for lower prices by increasing yields. The very limited area that has been taken out by Set-aside has often been of very moderate quality land, and any reduced production is then largely compensated for by higher yields on the better quality land which is left.

Nevertheless, many specialist cereal farmers are extremely worried by the prospect of still lower prices at a time when inflation is increasing and interest rates are high. As a result, many are seriously looking for alternative ways of using their land, either; by growing other crops such as beans, peas and oilseed rape; by putting it down to grass and keeping sheep; or by taking it out of farming altogether and putting it into recreational use, such as golf courses. It does seem probable that the days of many of the very specialised cereal farms are numbered, and that more diversified systems will take their place. This will help to placate those who have come to regard them as detrimental to the environment. But it would be quite wrong to allow too much of the land in Britain, which is so eminently suited to growing cereals in eastern and southern districts, to be removed from agriculture altogether at a time when the population of the world is still increasing at an alarming rate, and when the long term effects of possible climatic change are still so uncertain. A large strategic reserve of land available for the future, ready for grain growing on a large scale in the event of an emergency, simply must be maintained.

MIXED FARMS

Many of the changes that have taken place on all-grass farms and on purely arable farms are common to the more traditional type of mixed farm. Such farms vary, according to local soil, climate or topography, from units which are nearly all grass with only a very small proportion of arable land, to units which are almost completely arable, but which carry a small grass-based livestock enterprise, quite often a dairy herd. The tendency has been to simplify mixed farming systems as far as possible, and to cut the number of enterprises down to perhaps two or three, in order to reduce investment in machinery and buildings, which can be very high where a large number of different commodities are produced. This simplification has inevitably increased the size of the individual enterprises, as was illustrated on the all-grass farm, only in this case it also involves the intensification of both crop and livestock units.

The same pressures are there in each case: to try to increase output by the use of higher inputs of feed in the case of animals, or of fertilisers and sprays in the case of crops. The size of the individual livestock unit has increased enormously, whether it is a dairy herd, a sheep flock or a pig unit. Poultry production, on the other hand, has tended to move off the mixed farm altogether to be pursued instead by large commercial companies on small areas of land on a factory scale, even though there are still a number of sizeable battery units on larger farms. There are also quite a large number of very small units on mixed farms catering for 'free range farm gate' sales, or the 'natural' food market using organically produced feeds. But these only contribute quite a small number of eggs, compared with the overall market requirement.

As on grass farms, there has been a vast investment of capital in new buildings on mixed farms so as to bring them up to modern standards, both for the better winter housing of animals and the more efficient use of farm labour which was never possible in buildings erected in the eighteenth and nineteenth centuries. These were only built for the use of hand labour and not for machines. Modernisation also had to cater for the greatly increased size of the animal units. This has even involved the provision of sheds for the wintering of sheep flocks, which not only allow them to be kept off the land in the wet winter months from Christmas until March, but which also provide much better facilities for lambing, especially for those flocks lambing early in the year.

There has also been a large investment in buildings, both for the storage of grain and conserved forage, and for the protection of the very expensive cultivation and harvesting equipment required for the arable side of the farm. With grain yields some

These photographs illustrate a surprising paradox in modern farming. Sheep, which it might be thought spend their lives out of doors, are now housed to an increasing extent prior to lambing between January and March. Housing at this time prevents damage to pastures and helps to protect the sheep and their lambs in bad weather.

On the other hand, pigs, normally regarded as animals which are intensively housed for the whole of their lives, are now more often being kept outside on large arable mixed farms in light land districts. Each sow has its own hut, and a verandah to retain the baby pigs for the first week of life. The huts are moved to a fresh field annually and the land is ploughed to give high-yielding crops. Strong, healthy weaner pigs are produced under this system.

three times greater than they were thirty years ago, very large amounts may have to be stored after harvest. If good prices are to be realised, storage may be needed for up to seven months. Otherwise it may be necessary to accept a depressed price at a time when the market tends to be glutted after harvest. Adequate storage is especially important on the mixed farm because much of the grain is grown for direct consumption on the farm by the livestock through the following winter, and in the case of pigs or poultry, right up to the next harvest.

In common with the all-grass farm, the intensification of livestock production has brought with it problems of effluent and manure disposal, especially during the winter months (see pages 38 and 40). The situation is not usually quite so acute on the mixed farm as there may be arable fields which are scheduled for spring sowing with potatoes, barley or sugar beet. On the lighter soils on which these crops are usually grown it may be possible to use them for manure disposal up to, or even after, Christmas, without doing too much damage to soil structure from the passage of loaded vehicles. Most mixed farms produce more farmyard manure than grass farms because of the by-product straw from the cereal crops. Matured and well rotted manure can be spread and ploughed into the arable land after harvest in preparation for the next season's crop.

But this is not always as simple as it sounds. If the soil is at all heavy and sticky, a great deal of damage to the structure can be done by heavily loaded spreaders, especially near the field entrance in a wet autumn. It may then be very difficult to get good seed beds for a crop that should be planted by the end of October. The increased use of autumn sown barley has helped the farmer here. It is ready for harvest by the middle of July, or earlier, which leaves a longer period in which the manure can be spread on the stubble before the risk of heavy rain becomes acute. The introduction of low-pressure tyres for farm vehicles has helped to minimise damage to the soil in recent years, but even these can do harm on clay or clay loam soils if they are used injudiciously.

The slurry problems of the mixed farm are very similar to those of the grass farm, though there is rather greater flexibility because the slurry can be spread either on the arable or the grass, so making the timing of spreading less critical. But the problems of run-off and smell are just the same, though on the mixed farm, the risk of run-off into ditches from an arable field may be greater if slurry is spread onto hard ground after harvest and this is followed by heavy rain.

There is no question that farmers on all type of land and on all animal production systems are going to have to be much more careful about slurry disposal in future. The National Rivers Authority will be much stricter about the pollution of water supplies by either slurry, manure, or silage effluent, and this is as it should be. There is seldom any justification for the contamination of water courses by farm wastes of whatever origin, and the few who do offend only lay the whole farming industry open to criticism when the greatest part of it is quite blameless.

On most mixed farms, changes on the arable land have mirrored those on intensive cereal farms. In general, the introduction of highly mechanised silage making techniques has led to a very considerable reduction in the acreage of root and fodder crops, such as mangolds or kale, and to a concentration on grassland and ley production, using silage as the principal conserved feed for the winter months. This has led, in turn, to increases in field size to accommodate both arable and silage making machinery. But such increases in size have generally not been excessive, because many of the fields are also used for grazing cattle and sheep during the grass break in the rotation, and, as already stated, most farmers do not want very large fields for their grazing stock, even with the much larger herds that are common today.

The arable crops grown on mixed farms are generally cereals or oilseed rape, the cereals being divided between wheat to be sold as a cash crop, and barley used for home consumption. Only if the soil and climate are suitable are root crops, such as sugar beet and potatoes, grown to any extent, though this must be accepted as a somewhat broad generalisation, since it is not so true of Northern districts and of Scotland. Here, considerable areas of oats and swedes, more suitable for the moister and cooler climate, are still grown for livestock. Indeed, some of the oats are grown as a cash crop for sale off the farm as there is

Mixed farming: a healthy crop of potatoes with a new lake for irrigation and wildlife formed out of an area of unproductive, low-lying land.

still a good market for quality oats for both human and equine consumption. Mechanisation has largely removed the heavy labour cost of handling swedes which are very popular for feeding to both cattle and sheep in those areas where high yields are possible.

Another noticeable feature of the mixed farm today is the growing of what is known as a 'catch crop' of stubble turnips, rape, kale and perhaps forage radish, mainly for sheep grazing and fattening lambs in the autumn. This has become possible through the introduction of winter barley, harvested quite early in July. If the forage crop can then be sown directly, it leaves a good growing period before the days get too short in the autumn for much further growth. Very worthwhile yields can be obtained, provided that there is enough moisture in the ground at sowing to ensure adequate germination, and enough subsequent showers to keep the crop growing in August and September. This practice enables the land to carry two crops in the year, even though one is only a partial crop. It also allows fertility to build up from the manure left behind by the sheep which graze it and the roots which are ploughed back into the soil. But perhaps more important still nowadays, the forage crop mops up any surplus nitrates in the soil and prevents them from being washed down into watercourses in the autumn rains. So this recent development in cropping can have a beneficial environmental effect, and help in a small way to reduce levels of nitrate in surface water.

The environmental problems of the all-arable farm are far less evident on mixed farms. Most of the straw which is grown is needed for bedding or feeding to the livestock, so straw burning has never been widely practised. Hedgerows provide valuable protection for livestock, so not too many have been removed. Although very considerable changes *have* taken place on these farms, they are mainly in the nature of intensification rather than structural change. The majority are still farmed on fairly traditional lines, more especially those in the West and North. It is only on those mixed farms where livestock have been largely given up and there has been a major swing to cereal growing that there have been any problems.

MODERN PRODUCTION METHODS AND THE ENVIRONMENT

It will now have become clear that, as a result of all the changes that have been taking place in different types of farming over the past forty years, the whole appearance of the countryside must be very different from what it was at the end of the Second World War. Such changes must now be examined in relation to the rural environment as a whole, though first, it is important to try to define what the term 'environment' involves. It is a word, now very widely used — and sometimes abused — which can mean very different things to different people.

From the point of view of the countryside and of farming, there are essentially three different components which combine and interact to make up the so-called environment. These are physical components, biological components, and the soil itself, which contains elements of both. Under the heading of *physical* components can be included the size and shape of fields and buildings; the number and type of hedgerows; the presence or absence of trees; the colour of crops and grassland; and the way in which all of these blend in with the topography of the land.

Together, they can add up to a visual appearance, which to most onlookers should present a balance between the different elements, and the absence of excessive uniformity and dullness. Variety, after all, is said to be the spice of life, and there is no reason why this adage should not apply to visual sensations in the countryside.

Under the heading of physical factors should also be included the purity of the atmosphere and the absence of noxious gases, smells or pollutants such as smoke and dust, which can arise as a result of farming operations.

Then there are the *biological* factors. These include the number of plant and animal species present, both large and small, the presence of suitable habitats for them, and the importance of maintaining and even enhancing them, especially if they do not exist elsewhere.

Finally, there is the *soil*, not only its physical constitution and condition, but also its biological life. Almost as important as both of these is the effluent which drains out of it, since this can have significant effects well outside the immediate farming area.

These are three quite separate aspects of the environment in a farming context, even though they are very closely interrelated and integrated with one another. Each aspect can be of serious concern in varying degrees to particular sections of the community. Each one needs to be examined in a dispassionate way if the effects of farming methods are to be properly assessed in relation to the long term wellbeing of the countryside.

Physical Effects — The Visual Appearance

The chief criticisms levelled at modern trends in farming, and against cereal farming in particular, are that hedges are grubbed up, trees cut down, and fields greatly increased in size. As a result, it is said that in place of an attractive chequerboard of crops and grass of various colours and shades, there is a uniform monochrome expanse of land — green in spring and early summer, yellow at harvest, and brown in the autumn and early winter. There is now the added disadvantage that the very extensive area sown to winter barley, oilseed rape and winter wheat means that much of the land becomes green again very soon after harvest.

A further cause of complaint is that, with the greater quantities of nitrogenous fertiliser now being used on both crops and grassland, the green of the countryside is intense and uniform, instead of composed of a whole range of shades varying from near-yellow to brown, as was the case when less fertiliser was used. This is particularly evident with improved pastures in hill and upland areas.

Reference has already been made to the removal of hedgerows and trees and to the larger size of fields. There is no doubt that in *some* areas, the widespread removal of hedges and trees in the 1950s and 1960s, which led to much larger fields and to a more monotonous appearance, did radically alter the face of the countryside. Some farmers certainly overstepped the mark, and took out far more hedgerows than were strictly necessary in their desire to adapt the land to mechanised farming to save on increasingly expensive labour costs. There have not been anything like as many examples of this in recent years, in spite of claims to the contrary. Accurate statistics for the removal of hedges are difficult to obtain, since there is no official record of hedgerows. A recently published figure suggested that the rate of removal was as great in the 1980s as it had been in previous decades. But this is difficult to believe by anyone who travels widely through the countryside today, and who saw the scale of removal in the early 1960s. The only hedges that can now be seen being removed are on building sites or road construction schemes. One estimate has been made, based on aerial photographs taken during the last war compared with photographs taken today. Such a comparison could only relate to small areas for which comparable photographs could be obtained. Figures from such limited surveys extrapolated over the country as a whole would be subject to quite enormous errors as there is so much diversity both in types of hedge, in density of hedges in relation to land area, and in farming practice. The same criticism applies to very limited surveys on the ground in specific districts, since these can give very misleading figures when extended to cover the whole country.

In fact, a good deal of *replanting* of hedges is taking place, and some authorities believe that this is, to some extent, compensating for any old ones still being taken out, so that the position is really not now deteriorating at all. An example of what is happening today is shown by the survey carried out in the Stanton area of Suffolk. This covers 50 square miles (130 square kilometres) and 28 parishes, and it is a completely rural area with no towns. A tree and hedgerow survey was carried out first in 1984 and repeated in 1989. In the five years between the surveys, 63,718 trees were planted, one tree for just over every farmed hectare (2.4 acres) every year. In the last ten years 137,580 trees have been planted. Also in the five years, 2 miles of hedgerow were removed, but over 9 miles of new hedgerow established. As there are 380 miles of hedgerow in the area, there was in fact a net gain of nearly 2 per cent in hedgerow length. The area is largely arable with an average field size of 8 hectares (20 acres) and 15 per cent of the area is not cropped, being woodland, grassland, rough grazing, buildings, etc. So while it may not be possible to extrapolate from this for the rest of the country, these figures do give a picture of the position in an intensive arable area in Suffolk, which is often rather emotionally referred to as a 'prairie'.

Where hedges *are* removed today, it is almost always on smaller backward farms which have come under new ownership or management, and where some modifications are absolutely essential for the use of modern machinery. With the present inflated price of land and the high rents payable on it, a new occupier simply has to try to make the best use of his assets, and to use the limited labour efficiently. This inevitably involves the extensive use of machines which, in turn, necessitates fields of reasonable size in order to operate.

There has been a very noticeable change in farming opinion in the last decade in relation to hedgerow removal. This is partly due to public opinion, and partly to a realisation that it is not necessary, or even desirable, to have *very* big fields, either for arable or livestock farming, so long as they are a reasonable shape for mechanisation.

For arable farmers, a field size of some 20 hectares is now considered quite adequate to ensure efficient working. Some farmers are replanting hedges to improve shape and size,

An example of corner planting of woodland to provide cover for wildlife while making the shape of the field easier for working with large machines.

though not necessarily using old lines, while others are planting trees in small awkward areas to make fields a better shape for speedier working. The advent of the Farming and Wildlife Advisory Groups (FWAGS; see page 139) has done much to change the views of farmers on this and similar topics, and their advice has been and is being increasingly sought by those contemplating possible changes in field boundaries.

Many of the complaints about modern farming methods spoiling the physical environment only relate to quite a limited area, or even to a few cases in a particular district. They are not really applicable to farming on a countrywide scale, though of course those who have axes to grind, will try to make the public believe that they do. This is especially so with an issue such as hedgerow removal.

It *was* particularly bad in certain areas in the Eastern Counties and Midlands at the time when farm mechanisation was really intensifying between 1955 and 1970. But even then, it was of little or no importance in other parts of the country. It is certainly misleading to use figures derived from the past, as is often done, as if they were those relating to the present day.

Trees and Woodland

The felling of trees falls into a rather different category to hedgerow removal, as to some extent it is already covered by legislation. In the early days of rapid expansion in mechanisation, many hedgerow trees were cut down indiscriminately and often unnecessarily as a part of hedgerow removal. This era was quickly followed by Dutch Elm disease which led to a severe depletion of tree cover in some parts of the country. Small, often semi-derelict, patches of woodland and copses were also felled and cleared in order to bring the land into arable use. This was, again, of more consequence in some areas than in others. There is now strict legislation controlling the felling of trees and small woodlands, but, in any case, the lowered profitabililty of cereal growing has removed any incentive there might have been to clear trees in order to grow cereals. Nevertheless, there have been a few cases in the past ten years of owners felling without permission, and the strictest enforcement of the law is necessary in such cases. Anyone found guilty on this score should be banned from farming organisations, as the farming community should never be seen as condoning the flouting of legal restraints. Fortunately, such cases are very rare.

On the positive side, there are now Government schemes to encourage the planting of woodlands. There are also grants to assist farmers and landowners to replant where clear-felling is desirable because a particular area of woodland has reached the end of its natural life, and the only economic option is to cut it down and start again. Here, a condition of assistance is usually that the replanting should be of mixed species, with a combination of faster growing softwoods and slow growing hardwoods. Of course, clear-felling and replanting does cause some disruption to animal life, but this is of a temporary nature, and has none of the long term effects that occur when an area of woodland is cleared completely and brought into arable production. When one considers the amount of planting undertaken in the past, which provides the often well wooded countryside enjoyed today, there is a very strong case for giving as much encouragement as possible to landowners to plant quite large areas of trees. This is especially the case when there is surplus land available from growing food crops; when Britain

Establishing a new plantation on an area of poorer soil on an arable farm.
Spring onions are being harvested on the richer valley soil. The land rises
steeply to the Cotswolds in the background.

has an admirable climate for growing timber, and £5 billion
is spent annually on importing timber and wood products.

Though much play is made in certain sections of the press
and by some organisations about the wilful denuding of the
countryside of its trees and hedges, it is still quite normal to
drive hundreds of miles throughout Britain today, North and
South, East and West, without seeing a single example of
desecration: to observe a well ordered, well wooded, well hedged
or walled countryside, with an attractive blend of trees, hedges
and well-cared-for fields comprising a composite landscape of
mixed background, which is a pleasure to the eye. A more
relevant concern for the future could be the effect of the Set-
aside policy of taking land out of production which, if not
properly controlled could lead to derelict and weed-infested
fields. These could well become eyesores with little or no
beneficial effect on wildlife.

Farm Buildings

Another complaint concerning the appearance of the countryside centres on farm buildings (see Lowland Grassland Farming and the Environment, page 39). It is said that the old picturesque groups of farm buildings have disappeared, to be replaced or concealed by the erection of large, grey, concrete block-walled, asbestos-roofed, factory-type grain stores, cubicle houses and yards. Gone are the pleasant irregular shapes of the barns, and the nicely tiled roofs of the stables and the cattle sheds. In their place is something resembling a modern industrial estate which tends to stand out almost aggressively in the middle of the countryside. But the old type of farm building, particuarly those used for the housing of cattle, often with walls 1-2 foot thick, became hopelessly uneconomic to manage so far as the efficient use of labour and the handling of materials were concerned. They were often not very suitable for the livestock, being frequently very badly ventilated and unhygienic. As a result they have had to be replaced, in just the same way that industrial premises built during the Industrial Revolution have to be replaced by modern factories. It cannot be stressed too strongly that agriculture today is also an industry which has had to modernise its fabric and equipment in order to remain competitive in its costs of production.

Unfortunately, the cost of replacing old buildings using traditional materials is inordinately high. As a result, large buildings in which modern machines can operate, and which are capable of being adapted for a variety of different purposes as farming conditions change, have mostly had to be constructed from concrete and asbestos as the only possible economic alternatives. They may not look nice, but there really seems to be no choice.

Having said that, the siting of new farm buildings has not always been as carefully planned as it might have been and, in future, it might be more desirable for them to come under the same planning controls as industrial buildings. There seems little valid reason why they should be treated any differently, and it would help the image of farming in the eyes of the public if they were seen to come under similar regulations. It *is* often possible with good forward planning to make new buildings blend in with the countryside, and some judicious tree planting can go a long way towards hiding some of the larger ones. In practice, it is surprising how quickly new buildings on farms can weather and mellow, and after quite a few years, blend in well with their background, especially if the right sort of trees have been selected for the screening.

Farmers do find it rather difficult to accept some of the criticisms levelled at their new buildings when they see some of the monstrosities of new urban building, even on the fringes of the countryside itself. If farm buildings were to come under full planning controls, it is to be hoped that there would be a proper appreciation by planning committees and their officers of some of the difficulties faced by farmers in the siting of new buildings. There are cases where large intensive agricultural buildings, such as grain stores or pig and poultry units, on limited areas verge in scale on industrial buildings. The crucial question may then arise as to what is or is not an agricultural building. But at the end of the day, the important thing is that a new farm building should be sited so as to cause as little interference as possible with the general view of the countryside, that it should be designed to fit in with its surroundings, and, in terms of modern jargon, be 'environmentally friendly'.

Atmospheric Pollution

The final issue of importance relating to the physical aspects of the environment is the purity of the atmosphere. The burning of straw after harvest has given rise to one of the commonest complaints about modern farming methods. The effects have been particularly bad in dry, hot summers, though they are less of a nuisance in damp or wet years. The nuisance is caused in several ways; by huge clouds of smoke, by smell, by the fall of smuts and cereal debris, and even danger to the public.

Thick clouds of smoke blot out the sun, affect breathing and cause serious hazards to traffic, especially on busy motorways where accidents have been caused. Smuts descending from the sky when straw is burnt quickly under very dry conditions can damage paintwork, laundry, and horticultural crops such as cauliflower. The smell from a big straw burn can be very pungent, and the upward draughts created by the rush of air from a large fire can spread this widely to neighbouring villages or towns. But conversely, the burning of straw has been a very valuable practice for the larger cereal grower. He does not need straw for livestock feeding or bedding, and he often has far too small a labour force to be able to cope with baling, carting and stacking it for possible sale. Burning kills off many of the diseases carried over from one crop to the next in crop debris; it also kills most insect pests of the crop on or near the surface, which could harm the following crop. Its effect on the surface of the soil is very beneficial, especially on clays, as it leaves a layer of very friable soil which improves the structure.

All this, however, is now largely of academic interest. The disadvantages to the public have been deemed to outweigh the advantages to the farmer, and the practice of straw burning is to be discontinued by law after the harvest of 1992.

Smell is another contentious issue which has been growing in importance in recent years. The principal source of smell in the countryside is the manure from cattle, pig and sometimes poultry units. In all cases, the size of the *modern* unit means that both dung and urine have to be handled mechanically and stored in a semi-liquid state, whereas previously in small units and old fashioned buildings, it would have been absorbed into straw bedding and carted out once a year. Admittedly, there was a strong smell while this was being done, but it was only once a year, and country dwellers were brought up with it, and appeared to accept it as part of rural life. But now many of those who live in the country were not themselves brought up there having moved into villages and rural areas from towns where they were generally not used to smells. If they were, then they were not the potent, pungent smells of slurry from a recently opened pit. Pig slurry is the worst, though cow slurry which has been kept over the winter months can also be very offensive. The situation is not improved by the spreaders in the field throwing it up into the air to achieve a uniform distribution.

Various attempts have been made to alleviate this type of smell by adding deodorants or scents, but these are not really very successful. The only really successful way to avoid the problem is to separate the solid from the liquid as quickly as possible (see page 40), and to irrigate the liquid directly onto the field before the smell has had time to build up. But this is impracticable on many farms for a number of reasons, not least the cost of equipment, and so it is unlikely to provide a solution in the foreseeable future. It is certainly a difficult problem. The best course is for farmers to try to adjust their spreading to periods when the wind does not carry the smell into heavily built-up areas, and to plough it, irrigate it or inject it into the land as quickly as possible to minimise the nuisance. For their part, those people who choose to come and live in the country will have to be more prepared to make allowances for the problems faced by their farming neighbours who produce the milk they drink or the pork and bacon they eat.

Smells do, of course, arise on farms for a number of other reasons; sprays, for example, though generally speaking, these do not give much cause for complaint. What can cause a great deal of disquiet is the use of sprays in the vicinity of houses and gardens, especially if they are applied from the air by light aircraft. There is a danger of spray drift and damage to gardens if they are incorrectly applied, or if winds are too strong. The Food and Environment Protection Act 1988, in regard to spraying and the training of spray operators, should go a long way towards reducing such nuisances in future, but it is possible to foresee the day when aerial spraying may go the same way as straw burning, and become illegal.

There are, therefore, quite a number of essential aspects of modern farming which can have physical effects on the rural environment, giving cause for concern to conservationists, to visitors to the countryside, to planners, to those with interests in scientific fields, and not least, to those who live there. It is certainly true to say that most of the more extreme actions taken by farmers during the mechanisation of their farms are now a thing of the past, whatever claims may be made to the contrary. The great majority of those farming the land today not only have a much fuller appreciation of the importance of maintaining the appearance of the countryside as a whole, but are also very anxious to cooperate, as far as is compatible with making a living from the land they farm, with those who wish to preserve the essential features of an attractive, rural environment. Complying with the best principles of conservation costs farmers money. It is, therefore, most important that the profitability of farming should be held at a level which will still permit farmers to spend money in this way. If it is not, then rural areas will be the first to suffer, and the maintenance of an attractive physical environment will become infinitely more difficult.

It is only to be expected that there may always be a small minority who will not conform, expecially if financial pressures on farming become too acute. For them, stricter legislation may be the only answer. But this must be avoided if at all possible as it could have the effect of penalising those who are contributing positively at present, and perhaps making them less happy about setting examples in the future.

Biological Aspects of the Environment — Causes for Concern

Biological aspects interact very closely with physical aspects. The hedgerow is an excellent example here as it is important as a visual component of the countryside, and at the same time is a habitat for living plant and animal organisms. Another example is the draining of land, especially if that land is an important component of the visual landscape.

The principal causes for disquiet over the biological aspects of modern farming relate, in part, to the removal of trees and hedges, the clearing of watercourses and land drainage and, in part, to the greatly increased use of chemicals of one kind or another for various purposes, some of which might be inimical to wildlife.

The removal of hedges is claimed, with some justification, drastically to reduce the range of plant and animal species normally found in hedgerow bottoms, and also nesting sites for birds. To a greater extent, the same is true when small areas of woodland are cleared and the land brought into farming. Here, there is certain to be a loss of nesting sites and habitats for birds. But the clearing of one, normal, annually-cut hedge in the middle of a field should not significantly affect the bird life of an area, though this will obviously depend on what other hedges and habitats are left in the vicinity. The removal of high, thick, overgrown hedges will clearly diminish the number of

A wide, well-maintained hedge with well-established trees at intervals.
The wide hedge provides good nesting sites for birds.

available nesting sites and lead to some depopulation, but this may unfortunately be necessary for the passage of large machines such as forage or combine harvesters, while the shading effect of overgrown hedges can be detrimental to both crop and grass growth. Once more, it becomes an equation between the needs of wildlife and efficient production if the farmer is to make a living. A compromise should always be possible on larger farms, where a few higher hedges can perhaps be kept without interfering too much with the farming. Possibly better still, small areas awkward for working can be left to revert to scrub, so straightening out field boundaries and making them easier and quicker to work. On the small farm, land really is precious as there is so little of it and it is more difficult for the farmer to write any of it off without losing valuable income. Many larger farmers *are* now trying to provide more nesting sites for birds in a variety of ways.

Insect life and small rodent populations are obviously eliminated when a hedge is taken out, and to that extent, biological life is curtailed. Some insect species are beneficial to farming, while others, such as flea beetles and weevils are definitely harmful to certain crops, so it becomes difficult to draw up a proper balance sheet of gains and losses. On most farms the loss of a hedge will not seriously deplete the overall insect and small animal populations, though in the past this certainly occurred in areas where hedgerow removal was excessive. This was also true of the partridge, whose numbers declined quite dramatically, partly as a result of the removal of nesting habitats, and partly due to the reduction in weed populations and insect fauna through the use of pesticides right up to hedge bottoms.

The same factor has been largely responsible for the decline in butterfly and moth populations, which has been such a feature of the country scene in the past forty years. The loss here has been very significant, and is perhaps more apparent to the older countryside dweller, who can recall a far greater

Many farmers are now encouraging bird life in this way.

variety of species and much greater numbers of butterflies. The main cause has almost certainly been the elimination of flowering weed species in farm crops through the use of herbicides, flowers on which they relied for food at a very critical time in their life cycle. The removal of hedges, which provide suitable habitats for butterflies and moths probably compounded the loss as weeds would also have disappeared at the same time. But the main factor must have been the use of herbicides, for, except in a very few areas, there would still be enough hedges left to provide for good butterfly populations if that was the major controlling factor.

In recent years, increasing attention has been directed at field margins and their importance to agriculture, the landscape, wildlife and gamebird conservation. Farmers may be clear about the requirements of the crop, but less so about the best way to manage the field margin. The field boundary is, in many parts of the country, the only suitable permanent habitat for a wide range of animals; for example, about 230 species of

The wide grass headland and tall hedge provide excellent wildlife habitats, whilst the wheat crop is more uniform when not planted right up to the hedge.

insects and mites live on hawthorn, while hazel and dog rose each support over 100 species. Hedgerow birds, game species, mice, voles, butterflies and invertebrates all depend on field boundaries. Grey and red-legged partridges nest there almost entirely and up to 30 per cent of wild pheasants. The amount and quality of nest cover is critical to breeding and nesting success, as gamebird chicks need insect food for the first few weeks of life. Harmless insects that occur within the crop are depleted by insecticide sprays, and herbicides remove the weeds on which they live. A system of modified pesticide use on crop margins, developed by the Cereals and Gamebirds Research Project, controls some key weeds while leaving important insect host plants. This system has consistently increased average brood sizes of grey partridges and pheasants. Other groups of farmland wildlife also benefit from this technique, and species of wildflowers, which are now rarely seen, have been recorded in these 'conservation headlands'. Conservation headlands also act as a buffer, reducing spray drift into field boundaries, so benefitting the plants and animals living there.

A change from grassland to arable farming, even if it is not a complete change, usually reduces both plant and animal populations. Not only is there usually a tendency to reduce the number of hedges, but those on arable farms are generally kept more closely trimmed, and not allowed to become overgrown as may be the case on grass farms. They therefore provide poorer habitats for birds and insects. The question has to be asked: how important to the life of the countryside *as a whole* is this loss, primarily of insect and small animal life, and can it be ameliorated to any extent? It would appear that the loss of small animal life is not particularly serious, unless hedgerow removal is carried to more extreme lengths, but the effect on birdlife certainly can be. There may also be hidden interactions between species which have not yet become apparent. But experience gained from those farmers who have allowed a few hedges to grow up, or who have left small areas to revert, or who are leaving margins unsprayed or uncropped, suggests that it is possible to restore active wildlife populations quite rapidly, even in areas given over to intensive cereal crops. This almost certainly points the way to the future.

Effect on Plant Species

Only the effects of modern farming practices on animal and insect life have been considered so far, but there is concern, in some circumstances, about plant species as well. This is not related so much to the removal of hedges and woodland, even though this does reduce the range and number of wild plants, but to the effects of intensified grassland farming on the presence of rare species of plants which live in old pastures or adjoining hedgerows. Intensive fertiliser applications, the use of herbicides in controlled grazing systems, the drainage of wet fields or larger areas of marshland, and finally, the ploughing and reseeding of old permanent pasture, can all have very serious effects on the survival of rare plant species which only exist in those particular habitats. In the case of very rare plants, the problem may be solved by the designation of areas as Sites of Special Scientific Interest (SSSIs) (see page 136) which severely restricts the operations that a farmer can carry out on the site, and which may involve the payment of compensation for the resulting loss of income. The legislation introduced in 1986 relating to Environmentally Sensitive Areas (ESAs), whereby farmers may receive payment for following environmentally friendly systems of production, is also applicable to this sort of situation.

The improvement of moorland and hill pasture poses a similar problem, quite apart from the question of its scenic appearance. Hill improvement, involving ploughing up and lime and fertilizer applications, can also lead to a loss of rare species in certain circumstances, though in such cases this is often on a small scale, and a very large area of land usually remains on which such species can continue to thrive. So there is no question of an overall loss of species. Nevertheless, there are a few regions where large areas have been reclaimed in recent years, and where there is not a great deal of entirely natural pasture left.

There is a real dilemma here, and it is one of the more intractable problems associated with conservation and farming. On the one hand, the owner or tenant farmer has, at today's prices, a considerable sum of money invested in his land, even though it may be of poor quality. He has to ensure a reasonable level of output from the pasture if he is to get any return on his investment, and few, if any, can afford to play the role of the philanthropist. One of the quickest and surest ways of increasing output is to put the plough in, and reseed with productive grass species — assuming that the land is ploughable. If it is not, a considerable change can still be effected by surface cultivations, and a new seeds mixture. Both courses are drastic and will generally severely deplete many of the original, indigenous species of grasses and broad leaved plants which have become adapted to the conditions over centuries of acclimatisation.

The effect however may be of a temporary nature, for it is surprising how quickly many of the indigenous species regenerate and take over once more, if the management of the improved land is not kept at a very high standard. On the other hand, if the farmer is not allowed to drain, fertilise or plough the land, it may well become impossible for him to earn enough money to pay the rent or the interest on borrowed capital, since he cannot stock the land sufficiently heavily to sell enough animals from it. If only one field or small stretch of moorland is involved, and it is a large enough farm, the problem may not be very serious. Alternatively it may be possible to come to some agreement, through the ESA scheme, for the payment of a smallish compensatory sum, the farmer may be able to farm the land at a low rate of intensity, and still remain financially secure. On a smaller farm, the payments under such a scheme may not really be adequate, but at the same time, the area to be improved is probably relatively small also, in which case the improvement may not cause any serious biological damage, or spoil the scenery. But, if payment schemes are to become a widely adopted method for conserving existing conditions and maintaining wildlife habitats, farmers themselves must not be too greedy. They will have to accept that a slight loss in net income may be part of the responsibility they must carry as custodians of the land they farm.

The Wetland Areas

A comparable problem is met at the other end of the altitude scale — in low lying areas adjoining rivers or the sea. Notable cases in this category are the wetlands of Somerset and the Norfolk marshes, both of which have been in the forefront of the conservation controversies of recent years, though there are quite a number of other examples where improvement has caused disquiet on environmental grounds. It is not only the loss of rare species, adapted to the special conditions, which is involved here, but also the disappearance of a particular type of landscape, which is unique to the areas concerned. The farmer claims that he needs to be able to plough the land in order to get an adequate return from it. If this is done, and the land is planted to wheat, which often grows very well under these conditions (the land reclaimed on Romney Marsh many years ago is proof of this), the whole of the scenery undergoes a change. It becomes a large expanse of cereals, which looks, under the flat conditions, more like the American Middle West than the English countryside. Originally the marshes were stocked with cattle and sheep or cut for hay, and though this flat country may not be the most scenically beautiful in Britain, it does have a certain charm of its own, with the livestock and watercourses providing variety to the view. Of course, if the land when ploughed was farmed on a mixed farming system, there might not be so many grounds for complaint, but there is no guarantee that this would be the case.

In addition to the loss of valuable species of indigenous plants, which would follow ploughing and drainage, there is an additional dimension. This is the almost certain loss of animal species, particularly birds, both local and migratory, in the case of the Norfolk marshes, if the land were to come under the plough.

These areas provide classic examples of where ESA legislation can play a vital role, and fortunately, it already appears to have defused much of the conflict, where it has been applied. Under agreements made between the farmer and MAFF, the farmer will undertake to farm in an environmentally friendly way, will refrain from ploughing and reseeding, restrict fertiliser application and the use of herbicides, and generally agree reasonable levels of stocking for the land. Management thus reverts to a low/moderate input system, and the farmer has to accept lower returns than he might otherwise receive. In return, the Government will pay him an agreed sum annually, calculated on the basis of the difference between what *is* obtained, and what *might* have been obtained. Obviously, some hard bargaining is involved, but the agreements so far concluded appear to be working satisfactorily, and could well act as blueprints for the preservation of a large number of natural environments, which could otherwise be threatened by changes in management.

It is quite possible that the threat in future may not be so serious as it has been in the past, for at least two reasons. Firstly, because the surplus production of cereals, and lowered prices, will make it far less attractive to farmers to plough up old grassland, and there are no longer any grants available for land improvement, which previously made it an attractive proposition. In coming years, the costs of improvement may be greater than the anticipated returns, and the risks would not be worth taking. Secondly, because most farmers are becoming much more conservation minded, and more aware of public opinion, and will think twice before embarking on large scale reclamation schemes which would be politically unacceptable. The main incentives towards increased production, which once stimulated farmers to embark on land improvement, have now disappeared, and are unlikely to return, unless world shortages of food become acute or, for some strategic reason, Britain has to increase its levels of production again.

The change in attitude of many farmers towards conservation policies has been gathering strength for some years, largely because of the bad press that the farming industry has received. Many farmers have also come to appreciate that some of the methods used in the days of expansion after the war may have been detrimental to the environment in some respects, and that as guardians of the land, they have a responsibility to see that irreparable damage is not done. The Farmers' Unions, to their credit, have taken a strong lead here, and with progressive and enlightened farmers setting the pace in many areas, especially through the FWAG organisation, it does seem that there will be much more cooperation in the future with those who want to pursue active conservation policies. But it is important, in return, that the conservation lobby should not overplay its hand, since this could be counter-productive. Making impracticable and unreasonable demands, as some of the extremist pressure groups are prone to do, simply hardens attitudes against them, particularly when some of the arguments used are quite unsustainable. There will clearly have to be some degree of 'give and take' on both sides if the more intractable problems on both biological and physical fronts are to be solved.

The Effects of Pesticides

One of the most emotive issues in the biological field is the use of pesticides, that group of chemical substances that embraces herbicides, fungicides, insecticides, and some of the protective agents used in animal production. It is a long list, and one which grows year by year as new and more effective compounds are introduced to counter pests and diseases of crops and livestock. In fact, the increase in use since the first systemic herbicides for weed control were introduced in the middle of the 1940s has been enormous. It must be emphasised that strict controls are applied to the introduction of any new pesticide. All new compounds must receive a licence before they can be used on the farm, and they are subjected to comprehensive scientific tests, assessed by an expert committee of scientists who advise the Government.

Every release is controlled by the Food and Environment Protection Act 1985 and the Control of Pesticides Regulations 1986, and sanction has to be obtained from no fewer than six government departments. There is also a British Agrochemicals Inspection Scheme, which has been strengthened by the 1988 legislation, relating to the storage and use of pesticides on the farm. Under this, strict conditions are laid down regarding the storage of pesticides, and the disposal of wastes and used containers. In addition, it is obligatory for sprayer operatives to have a certificate of competence, which can only be obtained as a result of practical tests on the farm. Finally, all pesticides must come under fresh review after a period of ten years to ensure that there is no longer term evidence of deleterious effects.

The recent legislation was timely. There was certainly some lax handling of pesticides, dating back to the somewhat careless practices of the early days of pesticide use, and it was certainly time to apply more stringent standards.

But even if controls on the licensing and use of pesticides in the field have been tightened up, does the widespread use of this large range of chemical compounds have any effect on the biological environment? It must be admitted that there are no very clear-cut answers. Probably the most significant effect has been that of herbicides, mainly in arable crops, but also to some extent in grassland. Charlock, poppies, thistles, cornflowers and docks are seldom seen in cornfields today, (though they seem to return remarkably quickly if a strip or field is left unsprayed). In many grassland areas, too, buttercups are now seldom seen in pastures. Along with these more obvious species, a whole range of minor ones have almost disappeared from the more intensively farmed districts. With their disappearance has come a marked decrease in the butterfly and moth populations, which rely on the flowering weeds as a source of food at a vital point in their life cycles.

The use of insecticides has also probably contributed to the decline of other insect species, though their effect on the butterfly population has been relatively small compared with that of herbicides. In the early days of insecticide use, the long term effects of some of the compounds approved for use on farms were insufficiently appreciated: agriculture was not alone in underestimating the possible side effects of new pharmaceutical products. Dieldrin, for example, was to have an influence on biological populations in the wider environment far outside its original targets, and together with others in the same category, such as DDT, is now either banned completely or severely restricted in use.

Damage is always possible in the short term, for example, to bees through the spraying of oilseed rape or beans without proper notice to beekeepers, or to birds through leaving dressed seed corn on the soil surface of a field headland. In the longer term, delayed effects or a slow build-up of toxicity, causing permanent damage to populations of insects and larger animals, is probably very rare. Biological populations have a capacity for increasing very rapidly from a temporary set-back unless the breeding stock is reduced catastrophically. So far, this point of catastrophe does not appear to have been reached as a result of the increasingly widespread use of these compounds in farming, with the possible exception of the serious reduction in populations of butterflies and moths. It should be remembered that for every field which receives an insecticidal spray, there are many fields, both on the farm in question and on neighbouring farms, which do *not* receive it, and insect populations will be maintained on these unsprayed areas.

But that is not to say that an extremely strict watch should not be maintained on the effects of any new compounds which are introduced, or that constant monitoring of the possible long term effects of the older compounds should not be carried out. This was emphasised in 1989 when the Minister of Agriculture ordered a fresh review of the effects of a number of old, established pesticides.

It has to be appreciated that herbicides, fungicides and insecticides are now an integral part of farming, and that it would be impossible to maintain adequate food supplies without them in Britain, and especially in the world as a whole, where the increase in population is estimated at an extra three people per second (UN statistics). But two things are essential for the future. Firstly, the strictest watch must be maintained for any ill effects, both on the biological components of the environment and on the human population, i.e. the consumers of the crops and animals produced. Secondly, it needs to be instilled into all farmers that such compounds should only be used when absolutely necessary, and that it is obligatory that they are used only in accordance with the recommendations laid down for them. It is perhaps fortunate that the cost of pharmaceutical products is now very high, and could be one area in which farmers would be looking for economies in production costs. A further very encouraging factor in the overall use of these compounds is the development of new formulations which require much smaller amounts to be applied to achieve comparable effects. There is also the continued improvement in the design of sprayers. This will certainly mean that much smaller volumes of spray will be required in the future to get full coverage of the crop or weeds. The careless use of pesticides is now virtually a thing of the past, and although the overall quantities used are still high, especially for some crops such as fruit and vegetables, there are certain to be very considerable reductions in overall use as technology improves in the years ahead.

Effects of Modern Farming on the Soil

The final accusation in relation to modern farming methods is that, as far as arable farming is concerned, they are ruining the structure of the soil, bringing fears of future dust bowls, and that the quantities of chemicals used are not only poisoning the soil itself, but are also leading to an increasing risk of contamination of water supplies from excessive run-off into ditches and water courses.

Fears concerning soil structure are undoubtedly based on pre-war experience of continuous cropping with cereals, particularly in the United States, but also to some extent in Britain. But this experience was gained under completely different conditions from those of today. The Dust Bowl in the US was caused by a low input system practised on light textured, unstable soils in a dry climate after several drought years. The size of the crop grown under such conditions was quite inadequate to return enough organic matter to the soil to maintain its structure. With strong winds following a drought, it simply blew away. In this country, too, there was a certain loss of condition on some of the earlier farms experimenting with continuous cereal growing, for much the same reason. The crops were just not large enough to produce adequate root systems and stubble to maintain soil organic matter at a safe level. Light soils began to fall apart, while heavy land took on the consistency of putty when wet and of concrete when dry. Crops producing 15-18 cwts per acre (1.8-2.2 t/hectare) or less, which was standard for that time, were too small to provide the organic residues needed, and a vicious downward spiral of condition resulted, to the point where it was necessary to revert to grass breaks in an attempt to restore structure once more. The same thing happened during the Second World War under the Ploughing Up Campaign, and the need to produce maximum quantities of cereals to feed both the human and animal population. By 1943-44, signs of loss of structure began

Harvesting a high yielding crop of wheat with a combine capable of covering up to 25 acres (10 ha) a day. This crop grown with modern techniques certainly shows no signs of decreased yield from either disease or soil deterioration.

to appear, and some farmers had to be ordered to put land back to grass to remedy the situation.

But today the situation is completely different since, as we have seen, new technology in cereal growing has enabled yields to be more than trebled. Crops now have very extensive root systems below the ground, and a great deal of stubble is left above the ground by the combine harvester, which is returned to the soil (assuming that it is not burned). Though soil analyses show that organic matter levels do not rise appreciably under these conditions, but remain relatively stable, they certainly do not fall, in *most* circumstances, to a point where the structure becomes seriously affected. This statement may require some qualification as there appear to be *some* circumstances where continuous cropping can give rise to problems of structure. These are mainly on light soils, and on silts which tend to pack or run together in wet conditions. Quite serious soil erosion has appeared on some soils of this type in the West Midlands in recent years. It can be triggered off by the use of heavy tractors and machinery when the land is prepared for autumn sown crops of wheat and barley in wettish conditions. The wheels of the tractor leave quite marked depressions in the soil after the land has been worked down to produce a seed bed and the crop drilled. If these depressions run up and down the slope of a field, after heavy rain they can act as small runnels for the water that does not initially penetrate the surface owing to consolidation in the wheelings.

But as the volume of water increases, it begins to wash out the fine silt particles and what starts as a small runnel can quite quickly become a deeper rivulet, if heavy rain persists. Many tonnes of top soil and silt particles may be washed to the foot of the field, leaving the subsoil exposed in the depressions, which can become quite deep by the end of the winter.

Under such conditions, farmers will have to take much more care in future if they wish to follow intensive cereal rotations, and in more extreme cases it will be necessary to revert to grassland for a few years to restore the structure. If wheat is to be grown under these conditions, the tractor wheelings must either be cultivated out, or the cultivations and the sowing of the crop must be carried out across the slope, rather than up and down, as is the case in contour cultivation in other countries where erosion is a serious problem.

It is also difficult to maintain structure when growing spring sown barley continuously on light land. This is due to inadequate root growth in the short growing season, which is only some four months, and because light soils tend to dry out quickly in hot weather and as a result, there may be insufficient water available to allow for maximum growth of the root system. At the opposite end of the soil scale, difficulty can be experienced on very heavy clays which are inadequately drained, since here again, root growth may be restricted by waterlogging in wet years. But, by and large, there are now enough recorded examples of long runs of over twenty years of successful continuous cereal growing to prove that there need be no loss of soil condition, provided that the rules are followed and the standards of farming practice are high. If problems with this type of system do arise, they are more likely to be due to a build up of weed grasses, such as Black Grass, or Sterile Brome, than to a breakdown in soil condition. With evidence accumulating of resistance to herbicides by some weeds, this may become a worse problem in the future. But, at present, the good yields being obtained from the specialised cereal growing farms, on most of which there is a high input of both fertilisers and pesticides, it is quite safe to say that there cannot be anything much amiss with the soil conditions under which the crops are grown. If the soils were, in fact, in such bad physical shape, it would be quite impossible to achieve the results obtained today on many farms.

Are Soils Being Poisoned?

Exactly what is meant by the statement often made by those who strongly oppose modern farming practices that the chemicals used are in some way 'poisoning the land' is not clear. Objectively analysed, it must mean that the fertilisers applied, or the chemicals used for spraying the crops have harmful effects on soil organisms, or in some way render the soil medium toxic to the growth of plant roots. If that were the case, the soil would become denuded of the living organisms that assist in the decomposition and recycling of organic matter, in which circumstances, undecayed waste materials would accumulate. Alternatively, it might be argued that the use of some chemicals would kill off a certain type of organism thus upsetting the biological balance, and possibly leading to a build-up of other organisms harmful to root growth. This can happen in human or veterinary medicine where the removal of one type of bacteria can lead to an undesirable proliferation of another which may then become a problem in its own right.

There really does not seem to be any evidence whatsoever to support such fears. Under intensive farming conditions, the biological cycles seem, if anything, to be speeded up rather than impeded, probably because of the higher levels of nitrogen in the soil. This provides the soil organisms that break down the cellulose of waste plant matter, with the nitrogen needed for assimilation into protein and ensures the correct carbon: nitrogen ratio required for rapid decomposition.

If the fears were real, and the soil did become biologically 'dead', there would be a very rapid accumulation of undecomposed crop residues. But this is not occurring in practice, provided that the correct pH of the soil (acid/alkaline balance) is maintained, that soils are kept well aerated, and that waterlogging is avoided by efficient drainage. There is also no scientific evidence to indicate that the living systems in the soil are being deleteriously affected by the chemicals used in the form of herbicides, fungicides or even insecticides, most of which are very quickly broken down when they come into contact with the soil surface either by chemical or biological action. Even quite potent insecticides used to kill harmful soil organisms of a specific type, mainly in horticulture, do not appear to affect the balance of other organisms to any significant extent.

But, having said that, there are two areas for concern. These are firstly, the possible accumulation of a few chemicals which are *not* rapidly decomposed, and secondly, the risk of pesticide residues, particularly in fruit and vegetables, which might persist after harvest and affect consumers adversely.

A build-up of the few chemicals which are not rapidly broken down in the soil is likely to be of most concern where it begins to have deleterious effects on the succeeding crop. A number of herbicides can affect a crop plant adversely if the concentration is increased above a certain point, and yield can be quite seriously reduced. Such accumulation is most likely in very dry seasons, since the chemical tends to remain in the surface layers of the soil instead of being washed down and distributed widely throughout the depth of the soil. Breakdown by soil bacteria is also slowed down. There has been some evidence of this occurring with certain persistent herbicides, but how widespread the problem is, and what the level of residues needs to be with each herbicide, is not clear. It probably occurs very seldom in normal farm practice. But there *could* be a lesson here for farmers: sprays should only be used in the minimum concentrations necessary for effective action, and such compounds should not be used too often on the same fields, especially if there is any risk of a cumulative effect.

Are There Risks to Food?
The second risk—that chemical residues from spray materials might persist in or on fruit and vegetables, and might therefore

affect the consumer adversely—is of greater concern. Because of the large number of pests and plant diseases which attack fruit and vegetable crops, the level of spraying tends to be higher with them than in the average farm crop. Visual quality and uniformity of produce play such an important part in the marketing of both fruit and vegetables today that a grower simply cannot afford to produce blemished crops. This is especially so where extremely strict supermarket contracts are involved. Producers must therefore use a wide range of pesticides to ensure a saleable product at an economic price: they must also be sure, however, to abide by the maker's instructions in regard to the last permissable spraying date before harvesting the crop.

It would clearly be quite unacceptable that food for human consumption should be contaminated, either externally or internally, by chemical residues which might lead to metabolic disturbances or disease. The problem arises in defining what is a safe level for a residue of a particular chemical. There might, for example, be an infinitesimal presence in the produce of a compound used as a spray, but the amount could be too small to present any risk whatsoever to the consumer in relation to the quantity eaten. But if it was one which accumulated in the body, it might have to be banned, whatever the cost to the horticulturist or the farmer, in the same way that Dieldrin was banned from sheep dips due to its cumulative effect in the food chain. But if it was not cumulative, and only appeared in extremely small amounts, then the intake per day by the consumer would normally be well within the safety margin.

In the testing of spray chemicals, high concentrations are generally used in the diet of rats and if there are serious consequences, then the product will not receive a licence for field use. But, even if a high concentration in the diet of an animal the size of a rat did have an undesirable effect, it would not necessarily mean that a minute quantity, taken in from time to time by a human being would have a similarly bad effect so long as it did not accumulate in the tissues.

There is no such thing as a harmless chemical, only a harmless dose. This applies equally to natural and to synthetic chemicals. Pesticides are designed to kill weeds, pests and diseases, but it should not be inferred that they are general poisons.

Different countries have rather different standards for assessing residual pesticide risks and this can lead to some confusion, as a spray might be banned in one country, but not in another. Such an issue arose in 1989, with a spray used on apples, which was suddenly banned in the USA after a considerable period of use. It was not, however, banned in Britain, where the best scientific opinion held that, in the quantities consumed, any very small trace of residue which might be present did not constitute a health risk.

What will need very careful future monitoring is any possible contamination of fruit and vegetables imported into Britain from some of the countries in the European Community where spraying controls are not so well established. It would obviously be undesirable to prevent the use of a particular pesticide in Britain only to import large quantities of produce which had been liberally sprayed with it in the country of origin. With the freeing of markets in 1992, strenuous efforts will have to be made to synchronise standards throughout the Community. For the moment, the standards used in Britain appear to be as safe as can be reasonably expected, given the problem of deciding at what level of intake health might begin to be affected. Very large safety margins are always allowed in the calculations, and there is no evidence that dangerous residues have slipped through the net.

No discussion of the problems relating to the use of chemicals as pesticides in agriculture would be complete without a reference to the possibility of the contamination of water supplies through drainage. There is certainly a risk of this, as shown by the contaminaton of water in the United States some years ago and by the detection of very small quantities in samples of water in this country. Contamination could arise in two ways: firstly, by the flushing out of a persistent chemical in the soil by heavy rain after a dry period, or perhaps through heavy usage on a vegetable crop on a light textured soil; and secondly, by careless spraying or improper disposal of containers in which residues have been left. There have been occasional accidents in the past due to the latter cause, but there is no strong evidence that regular and serious pollution is occurring through overuse. The most likely geographical area to suffer would be the Fens and silt soils of East Anglia, with their very intensive area of vegetable and salad crops. With the tightening up of regulations in respect to water contamination, and increased refinements in chemical analysis, it is possible that residues will be detected more easily than in the past. Against that possibility is the fact that actual amounts of active chemical ingredients now being applied, and likely to be applied in the future, will decline, due to more efficient formulation and application techniques. The major water contamination problem of the future will more likely be that of nitrogen rather than pesticides, as the new legislation should prevent careless releases of pesticide in the vicinity of watercourses, and lead to their more efficient application to crops.

THE FUTURE — COMPROMISE AND MODERATION

If it can be accepted that the nation requires an efficient and productive farming industry on economic, strategic and social grounds, and that the best use of land should be made to provide the farmer with an adequate living and a fair return on his capital investment, can ways be found to avoid some of the more controversial aspects of modern systems of production? One of the difficulties in trying to find compromise solutions to specific problems of this kind is the tremendous variety of farming conditions and systems that exist in Britain, and this must be taken into account in any consideration of national policy. It would be only too easy to bring in regulations which might only apply, or could only be enforced, in a very limited set of circumstances, and at the same time penalise unfairly a large number of farmers in a quite different situation. Some assessment of the actual magnitude of the problems involved is therefore essential before any constructive solutions can be attempted.

To read some of the articles published, even in the more responsible journals, an average uninformed citizen might suppose that virtually the whole of Britain had become a Middle West prairie, mono cropped with cereals, with sterile soils washing or blowing away, with rivers polluted by agricultural effluents, and most of the wildlife eliminated. This is, of course, complete nonsense, and irresponsible nonsense at that, since the writers and editors of such articles must know that it is not a true picture which they are presenting. To take the 'prairie' statement as an example, the area under cereals increased by approximately one million hectares between 1946 and the late 1970s, and has been virtually stable, with occasional variations, ever since. This increase amounted to only some 8 per cent of the *total area* of crops and grass grown in Britain, or only some 5.5 per cent if the area of rough grazing is taken into account. A fair proportion of that extra one million hectares came from those areas which have always been dominated by arable farming in any case, and it is only in relatively few parts of the country where there has been any marked swing to cereal production. But by concentrating on those few areas only, it is easy to give the impression that changes nationally are much greater than they are. Though oilseed rape, often grown in rotation with cereals, has increased quite markedly, the actual area of cereals is falling rather than increasing, and is likely to fall still further as the measures introduced to discourage cereal growing owing to potential EC surpluses, become more widely adopted. The introduction of oilseed rape has brought more variety of colour to the countryside, but even this does not satisfy some critics, who complain of its vivid colour. It is an interesting comment on the situation that one group of people will lament the disappearance of yellow charlock from cereal crops, and the absence of this pleasing splash of colour, while another group will complain about the yellow appearance of its sister crop, oilseed rape. Could they even be the same people?

It is perfectly true that in *some* areas there was a big swing to cereal production in the 1950s and 1960s — that as a result there was some despoiling of the countryside, and that the character of the landscape changed and is still different today from what it was before 1939. These areas could well be some of the first to change over again to different systems of farming if cereal production becomes relatively unprofitable through attempts to reduce production by price mechanisms or quotas.

They may even be involved in Set-aside and the planting of new woodlands, so that in the course of time, the features of the countryside will change once again. It certainly does not mean that the Government should attempt to introduce legislation to limit cereral cropping in those problem areas, or try to restrict changes in the size of fields, or in the nature of a farm. This could easily have unfair effects on farmers in other districts, and in any event could only be made effective by introducing a high level of bureaucracy and added expense to the taxpayer. The Government can only act effectively in the widest overall context through legislation covering schemes such as Set-aside, or those which promote the alternative uses of land. Problems in a few specialised areas simply have to be kept in perspective, and attempts made to find local answers, so that the majority of farms in the country as a whole are left free of controls. If the national cereal acreage does fall, then there may be little need left for concern, in any case. But it is also clear that if farmers are to avoid demands for controls over activities which affect the physical amenities of the countryside, they will have to pay more attention to the feelings of the public than has sometimes been the case in the past. They may have to be prepared to adopt, and adhere to, codes of practice or sets of voluntary guidelines if they introduce or follow farming practices which are likely to affect the appearance of the countryside.

A start along these lines was made with the Codes of Practice in relation to animal welfare, and also straw burning, and there seems to be no valid reason why similar codes should not be adopted to cover other aspects of farming of a controversial nature. Such codes would have to be drawn up in consultation with amenity and wildlife organisations with strong leads by the National Farmers' Unions and the Country Landowners' Association, in order to guarantee as satisfactory an adoption of them as possible. The old attitude, 'the land is mine, and I can do what I think fit with it' will need to change throughout the whole of the farming community, and not just among its more enlightened members.

What seems to annoy some critics of modern farming methods is the claim that farmers and landowners know what is best for the countryside, and that they would not do anything to spoil their heritage. This is certainly very often true, but unfortunately there have been a few cases where it was obvious that a farmer or landowner has not always known what was the best way to treat his land and where excessive hedge removal, cutting of trees, or failure to maintain the fabric of the holding has led to a loss of amenity, without making any worthwhile return through increased yields or more efficient production. There is still some way to go before *all* farmers come up to the standards of the majority in the careful management of their land from the aspect of amenity value.

But if voluntary cooperation is to be achieved by the general education of farmers in the principles of conservation, and by the adoption of codes of practice, it is equally important that the general public, and especially that section of it which expresses its views most aggressively, should cooperate. They should in return be prepared to accept that the farmer is in business in a highly competitive field, that he has a great deal of capital at stake, and that he must be allowed to follow practices and adopt new technologies which will allow him to make reasonable profits. Such practices may on occasion necessitate operations which will bring change to the countryside; but change *is* inevitable and it is quite impossible to petrify a farm or an expanse of country in a fixed mould for all time. This has never been possible and it never will be. Conservation and protection lobbies will have to realise that it cannot be done, and that the best results will flow from a sensible partnership with those who are responsible for the day to day care of the land. If they are not prepared to accept an equally responsible attitude, then they cannot expect the farmer to meet them half way on those things that they would like to achieve in the field of conservation.

These principles apply particularly to the physical environment, but they are also highly relevant to those biological aspects which have already been discussed, and to the use of chemicals for different purposes in crop and animal production. Whether the 'natural food' enthusiasts like it or not, the farming industry will have to continue to use a wide range of chemical compounds if food is to be produced sufficiently cheaply, and in adequate quantity, to meet the needs of the vast majority of the population, and sold at a price which they can afford to pay. If yields fell and food had to be imported to make up a deficiency because British farmers were not allowed to use enough nitrogenous fertilisers, or because they were not allowed to spray their crops with fungicides, or because they were made to keep more of their livestock out-of-doors, it is quite certain that the food imported from other countries would itself have been produced under modern farming methods. So, in the long run, nothing whatever would have been gained from the attempt to tie the hands of the British farmer, and the nation would lose out financially from having to buy food which could have been better produced at home. The only winners would be farming competitors in France, Germany or the Netherlands, whose governments would, no doubt, be delighted to unload their surpluses in Britain.

But this certainly does not mean that if some producers are able to produce food by so-called 'organic' methods at a price which those who wish to eat it are prepared to pay, such methods should not be used. The question is one of supply and demand, and if the market is there at a price which some people can afford, then such methods of production will grow to meet that demand. But it would be quite wrong to say that *all* food should be produced in this way, especially if it means that poorer members of the community are unable to pay the price or that the producer is unable to make a reasonable living as a result of restrictions placed on his methods of production.

If it could be clearly and positively proved that food produced under some modern methods of production was *definitely* deleterious to the nation's health, or had seriously destructive environmental effects, then, of course, it would be necessary for the Government to step in and introduce legislation to prohibit the use of a particular compound or production technique, as it did with Dieldrin, DDT and certain animal feed additives. One of the problems in this field is that of isolating and proving exact effects and of arriving at some estimate of their importance — if, in fact, they are important at all.

Sweeping statements that artificial fertilisers are harmful to the soil, and to the person who eats the food grown on that soil, without any scientific evidence to support the statement, do not advance the cause of objective analysis. If such statements were true, it would be pertinent to ask why life expectancy has improved so dramatically since the middle years of the last century when Sir John Lawes initiated his first fertiliser experiments at Rothamsted. In practice, if fertiliser (and also probably fungicides and insecticides) were not used in agriculture, the present population of the world could not be fed. Certainly at the present rate of increase the world could not be fed by the year 2000.

So is there a middle road between those who wish to see the use of chemicals restricted and those who wish to use them in food production? If there is, it will have to lie in the area of moderate rather than extravagant use on the part of the farmer, and in a recognition by the opponents of chemicals that they are necessary to some degree if the producer is to make a living and the world is to be adequately fed. There has been a tendency for farmers in the past to regard some chemical sprays as an insurance against *possible* loss, and to apply them *in case* something might happen, rather than to wait and see whether they are really essential. Similarly, there may have been an over-use of fertilisers for some crops in an attempt to squeeze out the last increment of yield. The trouble is that the farmer never knows exactly what his crop actually requires in the form of plant nutrients, nor does he know at application time what the climatic conditions are going to be for the rest of the growing season, so he tends to play safe and in some years puts on more than is actually needed. With much tighter profit margins, which seem inevitable in the years ahead, this is an area where savings could well be made, and it seems probable that extravagant applications of fertiliser or pharmaceutical products will diminish, and that more moderate input/output systems will prevail. This is especially likely in the case of nitrogenous fertilisers, where compulsory restrictions on use may be imposed in those areas where water supplies are particularly at risk from pollution with nitrate residues. But it should be part of Government policy to monitor closely in the field the longer term effect of those compounds which might present hazards. This could mean allocating more funds to research, though regrettably this appears unlikely in the light of the attitude of the Government towards agricultural research.

There is really no sound and supportable case at all for the banning of chemical aids in agricultural production provided there is no evidence of harm to health. But all farmers should bear in mind that their use in large amounts *might* have undesirable effects, and they should always question whether a particular application is *really* necessary, or is likely to be financially justifiable. It would seem that the maxim for the future should be 'moderation in all things'. If it could be seen by the general public that this was a policy being pursued by farmers as a whole, it might go a considerable way towards allaying anxieties. Unfortunately, it would never satisfy those extremists who appear to be pathologically opposed to artificial aids in agriculture of any kind. It is time that the supporters of the 'Back to Nature' school began to appreciate the hard economic realities, both for farmers and the country as a whole, of the often vague and impractical nature of the theories which they support.

It has to be realised that production methods also have to take into account the problems of distribution, storage and shelf-life which are involved in supplying a predominantly urban population with its food.

RESPONSIBILITIES

If the agricultural industry is to play its fullest possible role in national life, there are responsibilities which should be recognised both by those who own and farm the land, and equally by those members of the public who are concerned in various ways about the land's well-being. There has been a very marked change in the attitude of informed farmers to the concerns and views of the public over the last twenty years — as is evidenced by the remarkable growth in the FWAG movement (see page 139). But there are still some farmers and landowners who resist change, and who appear resentful of what they see as a challenge to their right to farm their land as they see fit. This is, perhaps, understandable, when viewed in the light of some of the extreme attitudes taken up by their opponents; but it is not always helpful to those who seek to reconcile conflicting opinions.

The areas of greatest importance in the future are likely to be freer access to the countryside (see Recreation and the Countryside), the use of a wide range of aids to production, and the systems under which some farm animals are kept. Much of the criticism with regard to animal welfare is both uninformed and unfair — that relating to the indoor housing of sows for example. No one who has seen sows wading through deep mud to get to their food in an outdoor paddock on a bitterly cold, wet winter's day could possibly support a widespread adoption of outdoor pig breeding, except under low rainfall/light soil conditions, and the same applies to outdoor poultry.

But there are certainly still some systems which are justifiably open to criticism in regard to animal welfare — tethered sows in crates and overcrowded battery cages are cases in point, and such systems need to be phased out as soon as possible. Such changes, however, require considerable capital investment, and cannot be achieved overnight at a time when farm incomes are falling at a rate which leaves little or no surplus available for investment in new equipment. The further that profitability in farming declines, the more difficult it will be to bring about beneficial changes in production systems.

Care must also be taken not to phase out unacceptable options in the UK only to find that products from the same option are imported from other countries. For example, veal production from calves confined to crates is banned in the UK but veal is imported from Holland where the system is still used.

Farmers should always try to avoid laying the industry open to criticism in matters such as obstructing rights of way, the careless misuse of fertilisers and pesticides, or the pollution of watercourses, which are all contentious issues. Poor practice gets the industry a bad name and provides additional ready-to-hand ammunition for those who delight in sniping at farming; in many cases, the abuses might have been avoided with rather more care and forethought.

But if farming has its responsibilities, so has the public. It should realise that the farmer has both his own living to make and also that of his employees; on many smaller farms this is a real struggle under present day conditions (and when has it not been so for the smaller farmer?). The public should appreciate, too, that many of the changes that are advocated for improving the countryside will cost money, and that if things get too tight, survival becomes the rule of the day for the farmer, rather than trying to meet the demands of what often appear to be strident pressure groups.

The farmer and the landowner will always be the guardians of the countryside, and if such guardianship is to be exercised in a responsible, positive, and friendly way, not only money, but also some appreciation of the problems involved will be required. In the nature of his calling, every farmer is an individualist who does not like to be dictated to, or driven against his will. But he does have the best interests of the countryside at heart, and he will be extremely cooperative if approached in the right way, as will become clear in the following sections of this book.

Conservation — Its Principles and Practice

Having considered the background to land use in Britain, the development of modern farming systems and some of the problems they have brought in their wake, it is now time to look more deeply into what the non-farming 97 per cent of the population actually *wants* from the countryside. Essentially, these needs can be grouped under two main headings — conservation and recreation — though there are naturally very close links between the two.

Conservation is nothing new, but like the 'environment', it is a fashionable word which comes readily to people's lips. Unfortunately, it now conjures up other words like conflict and confrontation whereas compromise and common sense would be far more appropriate and helpful.

Conservation cannot mean complete preservation of the rural environment since, as we have seen, change is inevitable and it is quite impossible to freeze the countryside in a set mould for all time. As new technologies develop, established features which have become familiar (and which, in their time, were probably themselves man-made) will disappear and be replaced by something new. With an increasing population and shrinking resources, this is unavoidable, but it then becomes all the more essential that sensible partnerships, such as are now emerging between conservation groups and those who are responsible for the management of the land, should play an increasingly important part in conserving what is best in the countryside.

The Exploitation of Natural Resources

Since man first began to cultivate the land, he has dominated nature through his exploitation of its resources, often to a point where permanent damage and depletion have occurred. Some resources are non-renewable and, when they have been exhausted, can only be replaced by substitutes. Others are infinitely renewable if correctly managed in harmony with the soils and water on which they depend. If human populations grow too large or are forced to occupy harsh and unproductive environments, they can soon outstrip the capacity of the land to support them, as is only too apparent from recent experiences in some developing countries. Conservation seeks to relieve and to avoid such problems and to introduce good housekeeping principles into the management of the available resources.

Land Use in the UK

Land use in the UK has been of a generously conserving kind, helped by a benign climate. But, as we have seen, Britain is a heavily, even an over-populated, country sustaining fifty-seven million people on 240,000 square kilometres of land. Evenly distributed, this amounts to 0.4 hectare per person and not all of it high class land. The high standard of living which the population demands cannot be achieved without considerable pressure on our natural assets. Conservation problems are therefore becoming more acute, and call for a contribution not only from official and non-official organisations, but also from individuals concerned with wildlife and physical features.

Modern society, despite the gloomy picture sometimes painted by its critics, has brought leisure and easy transport to the vast majority. Increasingly, as we have seen, people turn to the out-of-doors for interest, inspiration and enjoyment. The pursuit of knowledge finds scope in the natural world; nature is thus a resource of a non-material kind. The conservation of nature for this purpose has to be set alongside its value as a material asset, although the two strands are completely interwoven.

Enthusiasm for nature is now a major concern of society and the principles of conservation, its needs and practice are more generally appreciated both by farmers and the public. But wider uses of the natural environment create an adverse impact on both its wild flora and fauna and on its physical features described in earlier chapters. It is the depletion of this heritage which is causing increasing anxiety. The rich heritage of our countryside owes much to those who managed the land in the past, and its future conservation must continue to be dependent on the sympathetic attitude of all land users. There is concern that policies designed to meet shortages which we may no longer face, together with economic forces, will put increasing pressure on the managers of land and other natural resources, and could encourage practices inimical to the conservation of nature.

ECOLOGICAL CHANGE AND THE MANAGEMENT OF LAND

During the last two thousand years very large ecological changes have been brought about in the UK. Primeval forest and swamps have been removed and the lowlands divided between farmland and an urban environment. The uplands are now composed mainly of treeless rough grazing. Most of the woodlands are artificial or at best semi-natural. The concept of protecting some of this original habitat goes back a long way. Hunting sanctuaries established by Norman kings still remain as important wildlife areas today, while the game preserves and fox covers of the eighteenth and nineteenth centuries have helped to maintain scattered woodlands as a long term economic investment. The management of Northern moorlands for game, combined with the grazing of sheep, developed a landscape and habitat which is enjoyed today.

There also developed during this period the concern and the skills needed to manage the land so as to maintain its productivity, both for the immediate future and for generations to come. Traditions of care and stewardship evolved, and land came to be regarded as a vital resource: in respect of its crops, livestock and timber and of those who lived and worked on it. Conservation became a fundamental principle of land use long before the word was used in its present sense, and this concept included wild animals, birds and fish as they represented a valuable source of food, game and sport.

The Modern Concept of Conservation

Until land began to be managed from 1900 onwards with a conscious intent to protect wildlife, our heritage of nature survived mostly through accident rather than by design. The transformation of the land over two-thousand years left and created a rich range of habitats, wild flora and fauna whose presence was largely incidental to the purposes for which the land was managed. The recent wish to protect and sustain populations of wild plants and animals as a scientific and aesthetic resource is really a modern facet of conservation.

LOSSES AND GAINS

Those who criticise farming and forestry interests are concerned really about the scale of loss and damage to wildlife and its habitats which has taken place since 1949. It is true that there have been some gains, too, in the creation of reservoirs, flooded gravel pits, sand and clay workings, new coniferous forests and tree planting programmes in the lowlands. But the more important losses make up a quite formidable list:

- Lowland grasslands such as herb-rich hay meadows where 95 per cent now lack a significant wildlife interest and only 3 per cent are undamaged by intensive agriculture.
- Lowland grasslands or sheep walks on limestone and chalk, of which 80 per cent have been affected either by conversion to arable cropping, by grassland improvement or by degeneration to scrub because of lack of grazing.
- Lowland heaths on acid soils, 40 per cent of which have been lost through conversion to arable cropping or improved grassland and afforestation, and some of which are now covered with scrub through lack of grazing.
- Ancient lowland woods of native broadleaved trees have suffered a 30-50 per cent loss by conversion to conifer plantations or to provide more farmland.
- Hedges have been removed on a large scale (see Physical Effects — the Visual Appearance, page 68), the loss varying greatly geographically, but being greater in the cereal growing areas of Eastern England and least in the stock rearing districts of the West and North. Out of approximately 500,000 miles (804,500 kilometres) of hedges in England and Wales, some 140,000 miles (225,260 kilometres) had been removed by 1974. Since then, the rate of loss has rapidly declined and there has even been some replanting.
- Lowland fens and mires — 50 per cent lost or damaged through drainage or chemical enrichment of drainage water.
- As a result of these changes, the flora and fauna have suffered

losses mainly due to the loss of habitat, especially those of the lowland grasslands, heaths and wetlands. Butterflies and dragonflies have shown the biggest losses as these insects are found in those habitats most strongly affected by modern trends in land use and appear to be particularly sensitive to change (see Biological Aspects of the Environment — Causes for Concern, page 77).

● Birds, on the whole, are fairly resilient and the populations of many species fluctuate more as a result of natural processes, especially the effects of weather. But they do show trends similar to other species where habitats have been modified. New upland afforestation, particularly in its early stages, encourages many animals, and especially birds. However, these gains have to be set against the losses of open moorland birds which include predators and waders. Bats have decreased and several species risk extinction. Problems of food supply, destruction of roosts and pollution have affected their numbers and all species are now legally protected.

● Botanical losses have been quite significant and out of 1,423 native flowering plants and ferns, 10 species have been lost completely and a further 149 have declined by at least 20 per cent since 1930. Of the 149, 69 belong to wetland habitats, 32 to permanent grassland, 18 to woodland and 14 to sandy or heathland habitats. Though, at first sight, these figures may well seem very light, it must be remembered that, once a species is lost, it can never be recreated.

THE RESPONSIBILITY OF THE RURAL COMMUNITY

The rural community of landowners, farmers and foresters determines land use over the greater part of this country and their influence on nature conservation is enormous. Farmers and foresters regard themselves as natural conservationists, a view upheld by the Country Landowners' Association and the National Farmers' Unions. Historically, landowners have shaped the face of rural Britain and left a legacy of wildlife and habitat which we enjoy today. During the growth of the nature conservation movement over the last hundred years, farmers and foresters have played a crucial role. The efforts to protect rare birds and their eggs, for example, has depended on cooperation from sympathetic landowners and farmers. Similarly, the expansion of nature reserves has been helped by the willingness of owners to enter into leases or management agreements, the new management usually depending on a continuation of previous practices.

Enthusiasm for field sports is also closely linked to the conservation of habitats important to wildlife. Much of the remaining semi-natural, broadleaved woodland has survived because of its value as cover for foxes, pheasants and woodcock. Grouse moors and deer forests in the hills are more important as wildlife areas than upland managed wholly for sheep or as conifer forests, while angling is an important factor in the conservation of lakes and rivers. Many farmers and foresters are keen naturalists and, as part of their own interests, look after the wild plants and animals on their land. Most feel a pride in knowing that they have something special and take particular care to cherish it.

Though this area of land appears unproductive, it provides grazing for sheep and an income from grouse shooting. The shooting butt in the foreground has been lowered to make it less conspicuous in the landscape.

In spite of this, as earlier sections of this book have shown, recent economic forces and policies for agriculture have led to some practices which do not favour nature conservation. The need for both maximum production and maximum crop protection is often incompatible with wildlife management, and good farm management today can be very different in its environmental effects from good management in the past.

The freedom and ability to practice nature conservation is strictly constrained by the economic necessity of modern farming techniques if reasonable profits are to be made to ensure the continuation of the farming business. Many farmers, however, will, if possible, refrain from carrying out operations which might be harmful to wildlife or habitats or will try to modify such operations. Others may even manage to create new habitats and improve the wildlife value of their farms.

CONSERVATION IN PRACTICE

There is now a much greater understanding about the importance of wildlife and landscape conservation, and the growth in the number of Farming and Wildlife Advisory Groups shows the value of bringing farming and conservation interests together. These groups encourage farmers, foresters and landowners to talk to conservation bodies such as county trusts for Nature Conservation, local branches of the Council for the Protection of Rural England and the Royal Society for the Protection of Birds, to discuss with them matters of mutual concern and to seek solutions.

Farmers are keen to receive advice about conservation practices which can be integrated with their farming operations and FWAGs can give such advice, often through their full-time Farm Conservation Adviser. Competitions, both local and national, increase interest in conservation practices. There is a FWAG or FFWAG (Farming, Forestry andWildlife Advisory Group) in every county in England, Wales and Scotland and one in Northern Ireland. The groups work in association with the Farming and Wildlife Trust.

Even allowing for the necessity to make reasonable profits, there is much that can be done on practically every farm,

A fertiliser spreader which can be tilted from the tractor driver's cab to prevent overlapping and spreading fertiliser into hedge bottoms.

however intensive the system may be. There will always be odd corners, steep banks, field and road boundaries, and wet places which cannot be farmed economically. These can be managed sympathetically for wildlife, and possibly planted with trees and shrubs where appropriate. For example, hedges and hedge banks provide a habitat for useful insect predators which help to control cereal aphids. Hedges and grass banks should not be sprayed and fertiliser should be kept out of hedge bottoms. Ponds and water courses are also valuable to wildlife and should be kept free from pollution and managed as sympathetically as possible.

Changes in the economics of farming, following EC measures to reduce the production of milk and cereals, are likely to result in land of lower potential being used for other crops or grass or even forestry. When profit margins are tighter, then land which has the potential to respond to inputs and produce high yields will be farmed intensively while land of lower potential will move to those crops or to grassland which can be farmed at a moderate level of production with lower inputs. Overall, these land use changes should benefit wildlife conservation as there will be less pressure to bring unfarmed land into cultivation and marginal areas will be farmed less intensively.

Control of Production
The accumulation of embarrassingly large supplies of agricultural commodities, with production well ahead of effective demand, led policy makers to seek ways to control production. Advances in agricultural technology have helped to produce a situation where the European Community is well able to meet its needs for the main commodities. Indeed, it has become an increasing exporter, which has helped to depress world markets. The cost of maintaining prices to EC producers has risen and Europe is in conflict with other major exporters. The EC has introduced measures designed to reduce the cost to taxpayers of agricultural support and to bring the production of the major crops more in line with demand.

The introduction of milk quotas in 1984 was the first step which marked the change from expansion to restraint on production. Other measures followed including *stabilisers,* introduced in 1988, and budgetary discipline, which limits the rate of growth of CAP expenditure to 74 per cent of the rate of growth of Community gross national product.

Milk quotas are a physical control over production, but the UK Government prefers action on prices rather than widespread use of quotas on other products. A mixture of price and supply controls and incentives for farmers to diversify are seen as providing the flexibility for the future. All these changes have taken place against a background of growing concern for the environment. As we have seen, public concern for the environment and about farming methods now exerts a major influence on agricultural policy. Competition between farmers and the general public over the quality of the countryside is of increasing concern, as is the move to country towns and villages by those who have never had any direct connection with farming or countryside pursuits.

Environmentally Sensitive Areas

Since 1986 UK agriculture ministers have been under a statutory requirement to maintain a reasonable balance between agricultural interests, the economic and social needs of rural areas, conservation and recreation. Policies have been introduced which are designed to encourage farmers to have more concern about the impact of farming practices on the natural flora and fauna. The Alternative Land Use in the Rural Economy (ALURE) package introduced in 1987 included woodland planting grants and funds for diversifying out of food production. The grants may be available to all farmers, as with Set-aside and the Farm Woodland Scheme, or they may be specific to defined areas as with the Environmentally Sensitive Areas scheme (ESA). Farmers in ESA designated areas are encouraged by financial incentives to farm in ways which are more friendly to the natural environment. The scheme is voluntary and has attracted a lot of interest. Measures are designed to maintain or recreate 'natural' features such as low input/low output, species-rich grassland through restrictions on the use of fertilisers and herbicides. In two ESAs financial incentives are offered to those farmers willing to adopt 'conservation headlands' based on the work of the Cereal and Gamebirds Research Project. The ESA proposals are clearly designed to encourage environmentally sensitive farming: this is the main objective of the scheme. Other schemes, designed to take land out of farming as the main objective, also aim to help wildlife and the environment. It is here that some confusion may arise because of the different objectives and the conflicts in management which can occur.

The ESAs contribute to reducing agricultural production. Lower inputs are an integral part of the scheme together with retaining land in extensive use rather than converting it to arable or heavily stocked grassland. Uptake under the scheme has been encouraging, but the ESAs cover only a small proportion of UK farmland; some areas are very small and the larger ones are located in marginal farming areas.

Set-aside

The major thrust to removing land from farming use lies in the Set-aside scheme and the Farm Woodland Scheme. Set-aside is, in the words of a former Minister of Agriculture, 'An opportunity for farmers to reappraise the way they manage their land during the next few years. Voluntary Set-aside is just one of a number of options open to you'.

The scheme is designed to reduce surpluses of arable crops. For taking out of production at least 20 per cent of land growing arable crops, based on the year 1987/88, annual compensation is payable up to £200 per hectare for five years. There must be no agricultural production from the land.

Options include fallow, which may be permanent, where the same land is taken out of production for the full five years, or rotational where the fallow area is moved round the farm as part of the arable rotation. The fallow can be whole fields, parts of fields or strips at last fifteen metres wide. The fallow option attracts the highest rate of grant, but there are restrictions on cultivations and the use of herbicides.

Other options include establishing plant cover, including game mixtures or green manure crops. Fertilisers cannot be applied to Set-aside land, although at the end of a one year fallow, preparations can be made for the next crop. Lime is allowed, but no slurry or farmyard manure can be applied.

Set-aside land can be used for non-agricultural purposes such as tourist facilities, caravans, riding schools or nature reserves but these attract a lower rate of grant.

The siting and subsequent management of Set-aside land can be very significant for wildlife. Land set aside alongside existing wildlife habitats will extend their value and provide a 'buffer zone' against farming operations. Set-aside land can also link existing features. Managing Set-aside land requires a different approach from that of managing existing farmland. To be of greatest value to wildlife management, it must be planned with this in mind. The Countryside Commission for England offers a Countryside Premium providing an incentive to farmers for positive management of land entered in the Set-aside scheme. At present this is only available in seven counties (Bedfordshire, Cambridgeshire, Essex, Hertfordshire, Norfolk, Suffolk and

Northamptonshire). It only applies to the permanent fallow option and covers such items as hedgerow management, encouragement of wildflowers and ground nesting birds, creating new meadowlands and creating grazing areas for Brent geese.

Set-aside land may only make a very small contribution to reducing total cereal production. It can however create serious problems for the farmer and his neighbours through the need to control weeds and pests. At present the problem is small and confined to a few areas, but if Set-aside becomes more popular then the scale will increase. There is no compensation for neighbours who, while not having set aside their land, may have to cope with problems from adjoining land.

One option under the scheme is to establish fifteen metre wide strips, either as headlands or across large fields. These provide access and can be attractive for gamebirds. Here it is essential to create permanent cover and cut frequently to prevent seeding of weeds and weed grasses.

Extensification

Reducing the level of fertiliser and pesticide use is another course which receives much favourable attention in some quarters. Lower inputs would reduce costs, lessen the amount of fertiliser and pesticide going into the environment, and reduce output. However, the formula ignores the fact that not all such inputs are used to increase output. Many are used to maintain the quality of the produce, to protect it from pests and diseases and to control weeds which may contaminate the crop so reducing its value, or even making it unsaleable.

Increasing concern about pesticides has led to proposals that restrictions should be introduced in order to limit the volume which is used. In Denmark, from January 1990, the total use of pesticides was decreased by 25 per cent, and it is to be reduced by a further 25 per cent by January 1997. Similar restrictions may be applied throughout the EC.

Herbicide use might be reduced if there was some way of calculating the threshold level so that farmers could decide whether weed control measures were necessary and, if so, at what level. Field experiments show that in cereals, herbicide

(Text continues on page 117)

Award-winning Examples of Farming and Countryside Conservation

The following farms, representing different farming systems from widely separated areas of the UK, have been chosen from among many hundreds that successfully combine farming with the conservation of wildlife and landscape. These farms demonstrate how enthusiastic farmers, working under very different conditions, have succeeded in marrying conservation with good, modern farming practice.

All the farms have received awards, some more than one, which recognise the contribution made to conservation and the continuing care for the countryside.

The *Country Life* Silver Lapwing Award, established in 1979, is given each year to the farmer who, in the opinion of the judges, has done most to conserve wildlife of all sorts on his or her farm within the constraints of successful commercial farming.

The Royal Association of British Dairy Farmers (RABDF) Landscape and Conservation Award is made every two years to the dairy farmer who, while managing a productive dairy herd, has successfully conserved landscape and wildlife features.

The Laurent-Perrier Champagne Award for Wild Game Conservation was created to encourage landowners and managers to create favourable conditions for wild game of all kinds and has been awarded in France for a number of years. More recently the award has been offered in the UK and has done much to interest estate and farm managers in the conservation of habitat for wild game which has also greatly benefited other wildlife.

The illustrations have been selected from the extensive collection of photographs taken over the years by Charles Jarvis, one of the judges of the Silver Lapwing Award, and from photographs by John Wilson and Eric Carter.

The authors are grateful to *Country Life*, the RABDF, Laurent-Perrier Champagne and to the individual farmers for permission to reproduce a brief outline of the award-winning farms.

1

IAN CRAWFORD
WESTER KINLOCH, BY BLAIRGOWRIE, PERTHSHIRE

Runner-up for the Silver Lapwing Award 1988 and joint third prize in the Laurent-Perrier Champagne Award for Wild Game Conservation 1990

Ian Crawford has been at Wester Kinloch since 1970. The farm of 510 acres slopes south so it is warm and has a relatively long growing season. It rises from 150 feet above sea level at the loch to 450 feet at the highest point. The soil is mostly light and gravelly.

It is a mixed arable and beef cattle rearing farm with 120 acres of long-term pasture on the steeper slopes and 110 acres of rotational grass which is cut for silage and then grazed. The 205 acres of cereals are mostly barley, and there are 25 acres of potatoes and 5 acres of turnips for winter feeding. The remaining 45 acres are small woods, bogs, roads, etc.

In 1970 the whole farm was grazed so that there was little natural regeneration. Since then fences have been replaced and realigned to protect all the wild areas, rough banks, bogs, the loch shore and woods from stock. Trees have been planted in hedgerows and rough areas with oak the main species, together with gean (wild cherry) and rowan. Ash and alder regenerate naturally. Three unobtrusive one-acre blocks have been planted with conifers to provide warm roosting for pheasants. Two plantations have been established, using the Farm Woodland Scheme, one 8 acres and one 5 acres, 65 per cent deciduous and 35 per cent conifer, with shrubs and hedges of mixed native species. These were planted on areas where arable crops did not pay.

The farm has some interesting old hedges which have been trimmed back, gapped up, double fenced and had hedgerow trees planted. This has proved a great asset in holding game, providing nesting cover and increasing insect and small bird populations.

Conservation headlands are used and have improved partridge and pheasant chick survival.

Up to 10,000 greylag geese have been recorded on the loch,

2

Beef cattle are an important farm enterprise.

Turnips are grown for winter feed. The farm rises to 450 feet above the loch.

3

Wild areas, rough banks and woods have been protected from stock.

New plantations have been established.

4

but 1,000 to 3,000 are more usual. Many ducks also use the loch in winter.

Ian Crawford is a shooting man with a keen interest in wildlife; his greatest pleasure is in the substantial stock of wild birds — over 120 species — identified on the farm since 1970.

A sound, commercial enterprise, this is a working farm where the farming has to pay for the conservation. A sign at the roadside proudly states: 'Conservation — Farmers do Care'.

Rides provide access to woods and encourage wildlife.

JOHN AND ROSEMARY BERRY
BILLINGSMOOR FARM, BUTTERLEIGH, CULLOMPTON, DEVON

Winner of the Silver Lapwing Award 1989 and the Royal Association of British Dairy Farmers Landscape and Conservation Award 1987

Billingsmoor Farm in the rolling Devon countryside comprises 232 acres of sandy clay loam, between 450 and 750 feet above sea level, rented from the Duchy of Cornwall. The main enterprise is the dairy of 112 Friesian cows; there are also a few pigs and poultry. The majority of the farm is grass for the dairy herd, either grazed or as silage. There are 30 acres of cereals (wheat, oats and barley) grown and marketed through the Organic Farmers and Growers Ltd. Ten acres of swedes for human consumption are not grown organically. A flying flock of around 130 ewes and lambs helps to keep the grass in good condition.

Two lakes and a number of ponds were created between 1974 and 1981, mainly in a rough valley of no use for farming and of little environmental value. Trees and shrubs have been planted round the ponds and the surrounding area planted with hardwoods and softwoods, creating a total conservation area of 15 acres. One lake is stocked with trout and the other with carp. Each spring frogs and toads breed; Canada geese nest and raise their young and the area supports numerous water-loving plants, insects, amphibians, wild flowers, butterflies and birds.

Odd field corners have been fenced off to make cultivations easier and planted with trees, and steep banks have been fenced off and planted with hardwoods and softwoods; three acres grow Norway spruce sold as Christmas trees. In all, 22,000 trees have been planted since 1979.

Hedges have been allowed to grow taller and thicker giving cover for wildlife, nesting for birds and shelter for livestock. Trimming is carried out in October and clumps of indigenous hardwoods left at intervals to improve the landscape.

Game feeders in the plantations are used by all the birds. Three different species of owl nest on the farm and owl nest boxes have been installed in two of the barns.

Conservation has always been an important priority, in

6

Lakes and ponds were created between 1974 and 1981.

Field corners have been fenced off to make cultivations easier and planted with trees.

(Above)
Steep banks have been planted with hardwoods and softwoods.

(Left)
Access has been improved for visitors.

8

conjunction with good husbandry, in the day-to-day management of this family farm.

Over 2,500 school children visit the farm each year and a GCSE teachers' pack is available.

Clear labelling provides information for visitors.

JOHN WILSON
MANOR FARM, IXWORTH THORPE, BURY ST EDMUNDS, SUFFOLK

Winner of the Laurent-Perrier Champagne Award for Wild Game Conservation 1989

Manor Farm comprises 1,680 acres, of which 1,292 are arable, 154 grazing, 110 woodland. The remaining 124 acres are farm tracks, ponds and water, amenity and conservation areas, and game crops.

The farm grows wheat and barley, sugar beet, peas and beans, and potatoes for seed production. About two-thirds is under cereals.

John Wilson has been at Manor Farm since 1962. Although some hedges were removed in the early days, as many hedges and boundary cover as possible have been retained. A cold, wet wood on the perimeter of the farm was removed and the land cropped. The lost area of woodland has been replaced by a number of new plantings at different sites around the farm. These have included shrubs and bushes in order to keep the covers warm in winter.

Hedges are cut regularly in alternate years, keeping them about 5 to 6 feet high. Care is taken to see that field margins and rough corners are protected from spray drift. Conservation headlands were introduced as an experiment in 1985 and today 75 per cent of the cereal acreage (20 out of 27 fields) is managed in this way. Breeding pairs of grey partridge have increased by 75 per cent and there have been consequent advantages for other wildlife, both plants and animals.

The farm has twenty-two pits or ponds which fill in the winter; at least eight hold water throughout most summers, and these attract wildfowl. In 1985 a new area of water was created and landscaped specifically for waterfowl which has led to a substantial improvement in the numbers of mallard, tufted duck, gadwall and other species. Teal and shoveller breed on the farm again now, as do snipe and redshank.

In 1983 hand-reared otters were successfully released into the Black Bourn river where the habitat has been managed and improved to provide low cover and areas for holts.

10

Barley and sugar beet with wide field margins and tall hedges retained to encourage game and wildlife.

New woodland designed to help game and wildlife.

11

A conservation headland in a wheat crop.

One of the new ponds with an observation hide.

Manor Farm is a commercial farm where shooting and conservation interests have been carefully combined to the benefit of the countryside and wildlife. Regular farm walks and open days are held to show the farming community and general public the results and to discuss the successes and occasional failures.

MICHAEL GOUGH
DUNCLENT FARM, KIDDERMINSTER, WORCS.

Twice runner-up for the Silver Lapwing Award and winner in 1988

Michael Gough has set out to make his 450 acres as attractive to wildlife as possible. Over 12,000 trees have been planted during the past ten years and nine ponds created or cleaned out. This is an excellent combination of conservation and farming; with the farm as the sole source of income it must be a highly efficient enterprise.

The farm policy is geared to sensible land management. Materials used to control weeds and pests have the least possible impact on wildlife and work is carried out carefully to ensure no over-spraying. Hedgerows are cut in very late winter so as to have a minimal effect on birds. A wide variety of trees has been planted in the hedges, and this not only enhances the wildlife interest but adds to the appearance of the landscape. A new conservation project is planned each year.

All the conservation projects are carried out by Michael Gough and his sole farm worker with other individuals or groups involved where possible. Many bird boxes have been erected on the farm by a young, local enthusiast. The suburbs of Kidderminster stretch to the farm boundary and Michael Gough encourages responsible public access. All the farm footpaths are walkable and all the stiles have been rebuilt. Many people, including school parties, are shown round.

Some of the 12,000 trees planted during the past ten years.

An odd corner planted with trees.

One of the nine ponds cleaned out during the past ten years. Water plays a vital part in wildlife conservation on farms.

Unploughed stubble and hedge trimmed to aid wildlife.

Farming and conservation are well integrated; for instance, stubble fields are left unploughed as a habitat for partridges and winter flocks of finches. The soil is light and subject to windblow so these fields will be treated with glyphosate and direct drilled with sugar beet, a system which benefits wildlife and demonstrates sensible crop husbandry.

Michael Gough is a keen shot and an excellent naturalist. The Dunclent Farm bird list covers over one hundred species.

Sugar beet (on the left) is an important crop on this farm. The grass banks provide a conservation area.

applications can often be reduced by one- to two-thirds of that recommended without any significant loss of efficacy and yield. But there are many problems in putting these concepts into practice — a weed population which, in spring, for example could be below the threshold likely to cause yield loss might, under later weather conditions, develop into a serious problem before harvest. To establish a threshold, weed population assessments are needed. These are expensive to carry out and, some, calculated on the basis of economics, are well above the level which would be visually acceptable. This is a concept worth pursuing, but its adoption will require a new style of farming and a different set of standards among farmers and those who advise them.

Farm Woodland Scheme
The Farm Woodland Scheme is confined to arable land and improved grassland. There is also a payment allocated for

Coppice regeneration on a Devon farm.

planting trees on unimproved land, including permanent pasture and rough grazing, in Less Favoured Areas. Payments will be made for up to forty years to farmers who convert agricultural land to woodland. Planting the whole holding is not acceptable; the lower limit is 3 hectares and the maximum is 40 hectares. Grants are paid each year for forty years where broadleaved trees are planted and 90 per cent or more consist of Sessile oak or European beech or a mixture of these, with shorter periods for mixed plantings. This scheme converts agricultural land to woodland; the trees must be managed to a satisfactory standard and this includes the necessary use of herbicides for weed control.

Large scale afforestation can have dramatic effects on the landscape and wildlife habitat, but small scale farm woodland is different. Many farm woods are established for shelter and can have profound effects in creating a better field environment for crops and stock and in protecting buildings. Mixed woods provide cover and warmth and in this respect are better than pure broadleaved stands. Increase in cover will benefit deer and a range of birds including game species.

There are a number of other grant schemes administered by the Ministry of Agriculture, Fisheries and Food, the Forestry Commission and the Nature Conservancy Council or through local authorities. These are aimed at encouraging wildlife and landscape works, usually on a modest scale; they are often discretionary.

It is worth remembering that what is good for the landscape is not necessarily good for wildlife, but improving wildlife habitats can often improve the landscape.

Protected Areas and Conservation through Persuasion

The number and extent of important areas now specifically protected for their nature conservation interest, such as nature reserves and Sites of Special Scientific Interest (the SSSIs noted earlier), provide a valuable nucleus of protected land. Within these areas, some of the best examples of natural or semi-natural vegetation with their distinctive plant and animal communities are protected by law. Outside these areas, the approach must be conservation through persuasion. The nature conservation movement has succeeded in raising the level of public awareness of the main issues, and, though the response has admittedly been variable, a trend of improvement has been established, and there is now a growing willingness on the part of farmers and landowners to consider the needs of nature conservation. The subject is no longer regarded with suspicion and is featured regularly in the farming press. Agricultural colleges include conservation in their courses and competitions and classes at agricultural shows have all played a part in making conservation part of the farming scene. Many farmers were 'green' well before the colour became so popular with politicians and the media.

Legislation and Personal Commitment

The Wildlife and Countryside Act of 1981 represented a major shift in the ability to safeguard important areas. That part of the law protecting flora and fauna has more effect with some groups than others, but the wildlife protection laws in general help to generate a climate of opinion which accepts the importance of, and responsibility for, wildlife.

The SSSIs and Nature Reserves are important, but if they are not to be left isolated and scattered over the farmed and forested countryside, they must be integrated with all the other conservation activities. Conservation is not just a matter of planting a few trees, digging a pond or maintaining the odd wood. To be carried out successfully it requires thought, care and long term planning. There is a symbiosis between farming and conservation, and farmers, managers and other users of natural resources must all be trusted to care for the land.

It cannot be repeated too often that such a commitment could never be properly implemented through the imposition of laws and regulations. It must arise from personal conviction and the fullest understanding and appreciation of all the complex issues involved. Only in this way will it be possible to achieve the ideal of a prosperous agricultural industry combined with a scenically attractive countryside available to all who wish to use it constructively for their recreation and enjoyment. Such an ideal will only become reality through a full partnership between all the interests involved.

Recreation and the Countryside

During the 1920s and 1930s, there was rapid increase in the urban population's use of the countryside for recreation. This produced both greater pressure on available resources and a demand for more access, in particular, access to relatively wild country, for example, the Pennines, the Lake District and parts of Wales, which were easily reached from the large urban centres of the North West.

Generally speaking, the average person experiences the countryside only when moving from one town to another or when using it for leisure and recreation; both of these are relatively new developments for most people. It is, of course, the availability of transport, either private or public, which has made the countryside accessible to a greater number of people where before it was the preserve only of the wealthy or of those who lived there.

The broad aspects of recreation probably affect the interests of a higher proportion of the population than conservation.

ESTABLISHMENT OF NATIONAL AND COUNTRY PARKS

The Scott Report of 1942 and the Dower Report of 1945 both dealt with land use and public access. The Scott Report recommended that the countryside should be used for agriculture and foresty, with industry located in towns, and that National Parks should be established in remote country areas of outstanding natural beauty. The Dower Report saw these parks as serving the whole community by preserving and enhancing the natural beauty of these remote rural areas. The report expected that the existing system of extensive sheep and cattle farming would be maintained, but with a much wider enjoyment of the area by the public through increased access, and that the land should remain in private hands.

Ten National Parks were created between 1949 and 1958 with the addition of the Broads in 1989, but the Dower Report had not anticipated the great increase in car ownership. The

National Parks have poor roads designed to serve a small local population and the great surge of visitors produces traffic jams and congestion, and erosion of the natural beauty they have come to enjoy. For example, over fourteen million people live within one hour's drive of the Lake District and the effect of only a small proportion arriving at the same time can be overwhelming.

The threat of destruction of well-known sites through overuse led the Countryside Commission, which was set up under the 1968 Countryside Act, to propose alternatives to the National Parks. One of these alternatives is the Country Parks, located near large urban centres, where special facilities have been provided for the parking of cars and for informal recreation. These areas are often referred to as 'honey-pots'. They are specifically designed to draw a large proportion of visitors away from the National Parks and the countryside in general, and so relieve pressure on them and reduce interference with their main function of agriculture and forestry.

Since 1968 the Countryside Commission has assisted in the creation of over 200 Country Parks, 240 picnic sites, 150 visitor centres and 629 kilometres of long distance routes (e.g. the Pennine Way) in addition to the 2,070 kilometres already in existence.

THE CLASH OF INTERESTS

The public desire for access, and the private use of land for farming or forestry, almost inevitably lead to a clash of interests and even to conflict. This is not confined to particular areas, although it has been most acute in the hills and uplands where the improvement of grazing by reseeding or surface treatment, and the conversion of open moor to forestry, have produced opposition both from amenity bodies and from naturalists. In fact, integrated forestry and improved grassland can lead to higher employment in upland areas, especially in the early stages when roads are being built and plantations established. Roads needed by foresters for the new plantations can be of great benefit to farmers by providing access to grazing land, and also to tourists who can enjoy the countryside.

Experience in Wales has shown that improved grazing and increased stocking rates following better access roads can more than compensate for the loss of land to trees. Lord Porchester (now Lord Carnarvon), reporting on the situation in Exmoor in 1977, recommended that, for areas of land which are regarded as most important for conservation and amenity, the owner should be compensated by central government for leaving the land in its present state, a concept subsequently incorporated into the Wildlife and Countryside Act of 1981.

The Number of Visitors

Visiting the countryside is the second most popular form of outdoor recreation after gardening. No other leisure activity outside the home can claim such a high rate of participation: 24 per cent of the population visit the countryside in a typical week, rising to 60 per cent in a summer month and 84 per cent over the year as a whole.

The Countryside Commission national survey of 1979 and that of subsequent years, shows that the impact of recreation on the countryside depends not only on the proportion of the population that visits, but also on how frequently the visits are made. The survey suggests that the majority who do visit, do so very frequently indeed. Over one-third had made at least 2 trips in the previous month, while one-quarter had made 3 or more trips with an average of 2.3 trips during each summer month. The survey has continued and in 1984 the monthly average of trips per person in each 4 week period was 4.5 with the highest figures of 6.0 and 5.9 in July and August respectively. The latest data for 1988 shows a fall in the monthly average to 3.7 trips with July and August at 5.0 and 2.9. This probably reflects the influence of weather rather than socio-economic factors, as 1987 and 1988 were rather dull and wet.

This represents a considerable movement of the urban population. On a typical Sunday, over ten million people may visit the countryside, and during a typical summer month, 100 million visits may be made by individuals. However, this does not represent a large section of the population, as much of the visiting is done by a small number of people making a large number of trips. In 1986, 49 per cent of those taking part in the survey made no trips to the countryside and of the remainder, 29 per cent were classed as casual users with 1 to 4 trips per month, 16 per cent as frequent users with 5 to 15 trips per month whilst 6 per cent made 16 trips or more per month. This suggests that active interest in the countryside is confined to a small proportion of the urban community.

The data allows profiles to be built up of different sorts of countryside users based upon frequency of use of the countryside:

- *Frequent users* making 1 or more trips a week account for 25 per cent of the population; they live in or near the countryside in good quality housing and own a car, and are more than likely to own a boat, caravan or horse.
- *Occasional users* make less than 1 trip a week and come from 50 per cent of the population; they are in clerical or skilled manual employment, situated three miles or more from the countryside.
- *People who go rarely or not at all* make an average of less than 2 trips a year and consist of 25 per cent of the population; they are mostly on low incomes, live several miles from the countryside and depend on public transport.

Countryside Commission support for the establishment of Country Parks and picnic areas has been of benefit to many millions of people, but has been of most advantage to those who visit the countryside frequently. *Local* areas are important, half of all trips taking place within ten miles of home. As most people live in urban areas, it is the country around towns and cities which is most visited and more than two-thirds of all trips were to fields, paths, woods, rivers and villages rather than to sites managed primarily for recreation.

Studies show that the value of visiting the country comes from a combination of the enjoyment of the intrinsic qualities that can readily be experienced in the countryside and the contrast with people's everyday lives. The study revealed an amazing richness of feeling and emotions, showing that the country matters to people from all walks of life and not just to the frequent visitor.

What People Look For and Expect

Recreation in the countryside clearly provides considerable enjoyment to those who experience it. Many people see the country as a necessary alternative to city life: a safe, healthy environment catering for relaxation and freedom from responsibility and the constraints of the town. But if they regard the countryside as an outlet — a total change to the experience of living in a densely populated area with limited open space — they also tend to look on it as an area of wilderness where they and their children can roam at will. Such a view must inevitably lead to misunderstanding, especially in lowland farming areas. That image of the countryside as a place of total freedom certainly accounts for at least some of the frictions. For a small minority, the countryside is a place where they can follow special interests — potholing, horseriding, fishing, serious hiking or rambling. But for the majority, it is mainly a place for casual activities such as driving, admiring the scenery, strolling around, or simply sitting in a car gazing at the view.

The survey suggests that over half the visitors to the countryside go to places that are not specifically managed for recreation. The use of actual agricultural land for recreation, however, appeared to be small since only 3 per cent of visitors visited farmed fields. Water, on the other hand, is a great attraction and 25 per cent of all trips were to locations where water was a principal feature of the environment. The survey also provided some evidence which helps to place the problem of access to the countryside in perspective. Just over one-third of the visitors went to areas specially managed for recreation or to villages, and this relieved the pressure on agriculture. Research shows the overwhelming popularity and importance of the wider countryside in comparison with the number of trips to Country Parks. The sheer scale of countryside visiting ensures that many Country Parks receive high visitor numbers and are an important part of local recreation provision, especially when accessible countryside is in short supply around towns. Recreation sites offer an experience which is distinct from that of the wider countryside in that they provide a wide variety of activities and represent an accessible and 'safe' form of countryside.

The available evidence shows that most visitors to the country are aware of and responsive to the rules of proper countryside behaviour and are able to enjoy themselves without causing any trouble. This is not to say that recreation does not bring problems, though more often through thoughtless behaviour due to ignorance than from actual malice. There seems to be a widespread, if ill-informed, support for countryside values, and a very genuine concern about its conservation, and this could perhaps provide a basis for more educational and interpretive work to improve behaviour. Attachment to the countryside is deeply embedded in most urban dwellers. But there is a lack of any real knowledge to support it, and much more needs to be done to foster a greater understanding and appreciation. There is a widespread desire for the countryside to remain natural. Special facilities are seen as being the opposite of what they envisage the countryside really is: not man-made, not planned and not organised.

The Future Impact of Recreation

As we have seen, recreation in the countryside involves a significant proportion of the population and some concern has been expressed about its future impact. Much of this concern can be discounted because the wide distribution of visitors limits its effect and, fortunately, the majority are well intentioned and well behaved. Concern has also been expressed about the possibility of a massive growth in countryside recreation. In the 1960s it was estimated that visits to the countryside were likely to treble by the end of the century. But, in fact, there was an increase in visits over the period 1964-77 of only some 8-10 per cent, and this was closely related to an increase in car ownership and real income. Between 1973 and 1979, the rate of growth actually levelled off or declined. Available data suggest that, though the actual number of trips made by individuals did not increase, the number of people making trips in an average summer month did increase from 36 per cent of the population to 43 per cent, a growth of just over 4 per cent a year between the two dates. By 1984 this figure had increased to 70 per cent, but fell away steadily to 60 per cent of the population making trips to the countryside during the summer months of 1988.

It is expected that the total population will rise by 2.6 per cent over the next ten years but there will be important changes in age and household structure. By the year 2000 there will have been a 15 per cent growth in the age group 30-44 which now represents the most frequent countryside users. There will be a fall of 20 per cent in the 16-29 age group who are also frequent countryside users. The number of retired people under 75 will fall by 10 per cent but there will be an increase of 20 per cent in those over 75, which will result in a slight decline in countryside use among older people. This will be more than offset by the continuing depopulation of the inner cities, with a likely further three to five million people housed in new towns or on the edge of existing towns and cities by the turn of the century. These people will live nearer the countryside than at present and will be more able and more likely to visit it.

Mobility will continue to increase, although car ownership is likely to rise only slowly, and public transport is predicted to decline. A rise in incomes and an increase in leisure time by about 5 per cent may well encourage more countryside visiting. Increased public awareness about, and concern for, conservation of the natural environment suggests that public enjoyment of an attractive countryside will grow in importance. Conservation policies must, therefore, play an important part in future recreation strategies.

Assessing future developments is difficult and the factors related to countryside recreation are complex. Income levels, car ownership, leisure time, economic and unemployment factors are all involved. The state of the economy, the price and availability of petrol, and changes in disposable leisure time must all influence recreational visits, and there may also be changes in the distribution of trips from weekends to weekdays. Although there is a potential for further growth, it may well come at a slower rate than in the past, and there could be a sizeable minority who might desire to visit but do not have the resources to do so. There are undoubtedly opportunities for making more and better provision near to urban areas, and for making access easier by improving public transport and by creating footpath access points. Country Parks provide all that is required by many people and they could well be expanded and promoted.

Towards a Greater Understanding

Whatever provision is made in the urban fringe, however, there will always be a considerable number of people who will want to continue to use the open country. Although their behaviour may occasionally leave a lot to be desired and can be harmful to farming, they are, on the whole, well intentioned, even if not always knowledgeable about countryside values. Conflict between visitors to the countryside and farmers could be reduced if more farmers were to accept that *most* people are normally well behaved. For those who are not, education and interpretation might help to reduce conflict. Existing programmes certainly need to be expanded, particularly in those parts of the country adjacent to larger centres of population.

It is, in fact, very encouraging to note the increasing number of teaching farms and farm centres that have sprung up and are now available for visits by the public, some with special facilities for viewing stock and watching farm operations, such as milking, in progress. Both college of agriculture farms and ordinary commercial holdings also attract large numbers of visitors to open days. These arouse a great deal of interest and many people are clearly keen to meet farmers face to face and to learn more about modern farming and its problems. Presented with the facts at first hand on the farm, very few remain critical. The British Food and Farming Year in 1989 brought together the whole food and farming industry to speak directly to the British public whose response demonstrated their interest. Nearly a million people attended the ASDA Festival of British Food and Farming in Hyde Park and many more visited local events. Teachers' resource packs were prepared and distributed. The success of the year emphasises the need to continue the dialogue on a permanent basis.

Another very interesting development has been the growth of 'pick your own' fruit and vegetables. This has brought a number of townspeople into direct contact with growing crops and often with those who produce them, even though the numbers using these farms declined from 32 per cent of those in a survey in 1984 to 21 per cent in 1986. Some growers have begun to provide interpretive facilities and this again helps to show interested customers something about modern farming practice. But much more could certainly still be done to provide those who are interested with first-hand knowledge of farming practice and problems.

Footpaths — A Special Problem

Footpaths, which once provided important lines of communication between outlying farms and villages and short cuts to work for farm employees, are now seen by visitors, and especially ramblers, as providing public access to the countryside. This often leads to conflict where footpaths, no longer used for their original purpose but kept open by ramblers and hikers, cross intensively cropped land or fields with grazing stock. There is an excellent opportunity for compromise on the part of the farmers, landowners and walkers, and it must not be missed. Well maintained, clearly signed paths, with no unnecessary obstructions, encourage users to stay on the path and give confidence to townspeople walking in a strange part of the country.

Those who use footpaths can help by agreeing more readily to the diversion of existing paths to the headlands of fields where they cause little interference with farming operations, and perhaps to the closure of those which are little used. There is a need for the provision of new paths where this can be shown to be necessary, especially in areas of scenic beauty. If farmers could appreciate that it is better to control the access of the public in such cases through the provision of a well sited and well signposted path, and, in some cases, even areas for car parking, they would experience much less interference from uncontrolled access over their fields.

The legal obligations regarding footpaths are fully explained in the code of practice, *Ploughing and Rights of Way*, published jointly by the Countryside Commission and the Ministry of Agriculture, Fisheries and Food.

That understanding and compromise can provide a satisfactory solution is illustrated by those landowners who,

A good example of facilities for walkers in the countryside, including a new car park at the centre of a footpath network. Controlled access and good signposting of paths can greatly improve relations between farmers and the public.

after negotiation and discussion, have arrived at a satisfactory reorganisation of the footpaths on their farms: something that is really essential when, as in some cases, every field on a farm may be crossed by a path, and sometimes by more than one.

Maintaining the Right Image

A countryside that is well farmed can be attractive and interesting provided that visitors have some knowledge of the agricultural practices they see, and have access to interpretive facilities. But to combine good farming with sympathetic wildlife and landscape conservation demands a commitment on the part of those who manage the land. Well-grown crops and well-managed livestock, within a framework of cared for hedges, trees and woodland, show that the countryside is fulfilling its basic function of providing food and other raw materials. Such a landscape will also fit the image that most visitors have of the countryside.

Reflections and Perspectives

This book has traced the recent development of the agricultural industry in the UK and the changes that have taken place in the ways in which crops and stock are produced — changes which have affected the countryside and the people who work in it, and which have often aroused considerable public concern. Change is inevitable and in the past, new farming methods have been developed and have created new landscapes and new systems. The enclosures at the end of the eighteenth and early nineteenth centuries were dramatic; they created social upheaval and upset many who had previously enjoyed the traditional open fields. Such people would probably feel very much at home today in those areas where larger fields were recreated twenty or thirty years ago.

Writing in 1832 John Clare expressed his feelings about what was being lost:

> Where bramble bushes grew and the daisy gemmed
>   in dew
> And the hills of silken grass like cushions to the
>   view,
> Where we threw the pismire crumbs when we'd
>   nothing else to do,
> All levelled like a desert by the never-weary plough,
>
> Enclosure like a Buonaparte let not a thing remain,
> It levelled every bush and tree and levelled every
>   hill
> And hung the moles for traitors — though the brook
>   is running still
> It runs a naked stream, cold and chill.

Some have expressed similar sentiments today.

The modern farmer supported by research, advice and education has, perhaps, been too successful and by using the new technology, is able to grow more of some basic foods than the country or the EC requires. But the nation needs food, and it is farmers who provide the raw materials with which the food industry keeps the supermarket shelves stocked with such a vast display of produce. So the debate rages about ways in which total production can be reduced without going so far as to produce shortages. In fact, production can never accurately balance demand as, despite all the modern aids, the weather has such an unpredictable effect on crop yields. Some surplus capacity has, therefore, to be built into the system to ensure stability and continuity.

The potential to produce large surpluses still exists, but is to some extent constrained by CAP policies of price restraint, production quotas and a stabiliser mechanism which determines levels of production. If pre-set levels of production are exceeded then the stabilisers initiate substantial falls in producer prices and thus reductions in production.

Food manufacturers and supermarket chains exert a considerable influence on producers and production systems, often laying down stringent standards and precise growing techniques. The consumer, quite rightly, demands high quality and produce free from damage caused by pests and diseases. Pesticides, as has been stressed more than once in this book, are not used just to make life easier for farmers and growers; they also ensure that produce is free from weed seeds which may be toxic and from the ravages of and presence of numerous insects, moulds and mites. With the bulk of the population in large towns and cities there must be some delay in transporting food to its final destination. Thus high standards of hygiene at all stages of production and 'sell by' dates are essential to help to ensure quality. But these can only be achieved through volume production and stringent controls.

There are some who argue for smaller family farms as a way of ensuring that the countryside is managed sympathetically for wildlife. But would small farms necessarily lead to a cosy landscape with a varied wildlife? If commodity prices fall, then it will be the smaller farmer who will be most affected and under pressure as he has a smaller income and less scope to reduce costs. His one competitive commodity is his own labour: he can work even harder to stay solvent. He will have less cash and less time to devote to conservation and will not be able to tolerate any losses through pests of any kind. Smaller farms would certainly bring more people back into the countryside but unless commodity prices rise, they and their families would be condemned to low incomes, hard work and drudgery.

Mechanisation has largely reduced the need for hard labour, and no one has ever settled the argument as to whether machinery drove out labour or machines were produced to do the work of those who left farming because of low wages and heavy work. Modern machines may not have the attraction of older methods for the visitor to the countryside, but they have reduced the sheer grinding hard work of the past. The speed with which operations can be completed also enables crops to be planted and harvested under the best conditions with fewer losses through bad weather, and less danger of damage to soil structure.

New crop varieties, systems of crop production, mechanical aids and animal feeding and health care cannot be 'uninvented'. There is no way in which — apart from some cataclysmic event — farming or any other industry can return to the past.

The farming industry is now facing restrictions on production and can no longer, as in the past, expect to meet rising costs by increasing output. Milk quotas have had a dramatic effect on dairy farmers, some of whom have gone out of business. Restricted output at a controlled price has led to changes in milk production methods: to the use of less concentrated feed and more reliance on grass and grass products; in fact, to a still more intensive use of grassland. Quotas have not been introduced for cereals but most growers expect lower prices which, together with the effects of stabilisers to reduce prices when total production reaches pre-set levels, can be expected to influence cereal growing. In the long run, this must result in cereals no longer being produced on land which is only marginally suitable. Then, only land which can sustain high yields will continue in cereal production. Inputs of fertilisers and pesticides will be used more carefully and may be reduced, but they will certainly not be cut out all together.

Some argue for a return to rotational farming with more short-term leys. But what use will be made of the extra grass? Milk production is restricted by quotas, while beef is overproduced and sheep, although reasonably buoyant, could easily go into overproduction as well. With a few exceptions, farming seems likely to remain broadly divided into arable cropping or grazing livestock. As we now have the potential to grow all the food we need (and more), it seems very likely that some land may go out of cultivation, as has happened in the past when demand dropped, and this will be in addition to land lost to new housing and to industry. With government encouragement, some land will be planted with trees for amenity and timber production. There is certainly no need to reclaim land for food production. Some marginal land, unless assisted by EC or government aid, could well revert to bracken or scrub or to forest and, even in the lowlands, some land which is less productive may go into woodland or into uses other than farming.

Concern about the nation's diet and possible effects on health has led to suggestions that less fat and sugar should be consumed, with fibre and fruit forming a greater part of the diet. While not accepted by all the authorities, there is a move to encourage these changes in eating habits. This could have a profound effect on farming, leading to changes in ways in which livestock are finished for the market and in less emphasis on high butterfat in milk. Overall, there may be a move away from 'red' meat — beef and lamb — to 'white' meat — poultry and pork — which are seen as containing less fat. Whatever changes there may be, they cannot happen overnight and time will be needed to adapt production systems to the new requirements.

There is also a small, but growing, market for health foods — often organically grown. While it is true that such farming methods use no artificial fertilisers or pesticides, the increase in their number is likely to be relatively small and specialised. Pure organic methods could not produce the volume of food needed for the large population of these islands; there would also be problems with some pests and diseases which could contaminate produce. There are some variations on the organic theme which allow the use of specified materials under particular conditions and these may also attract more interest.

It is necessary to retain a sense of perspective since we cannot go back and we must all eat. There are many people in the world who would regard our attempts to reduce food production as crazy. Some aspects of agricultural production will certainly need to be modified, but others are essential if we are to supply our own markets and also meet the needs of the consumer and the health authorities. We also need to be careful not to place restrictions on the home farmer which do not apply to overseas producers while allowing them to send food here. For example, there are rigorous tests at home on laying hens for salmonella and affected flocks are slaughtered, but imported eggs may come from countries which do not apply such regulations.

Changes in the Common Agricultural Policy which ultimately controls the pattern of farming, will not necessarily, or on their own, bring about more conservation. We will still need to produce a great deal of food and maintain emergency stockpiles. We also need to understand and evaluate apparently non-productive areas and to appreciate both agriculture and conservation as legitimate forms of land use.

It is nonsense to think of agriculture as being in opposition to conservation. Agriculture is by far the greatest and oldest application of ecology: it is man's earliest and most persistent effort to mould the environment to his advantage through ecological manipulation. It is a positive, all pervading application and an essential partnership with conservation. Indeed, farming and conservation have a symbiotic relationship. There need be and should be no conflict between agriculture and conservation. What is required is for those concerned with agriculture, conservation and ecology to understand each other and explain themselves to the general public, and for there to be a better understanding of the different viewpoints.

Appendices

THE WILDLIFE AND COUNTRYSIDE ACT 1981

This Act:

- repealed and re-enacted, with amendments, the Protection of Birds Acts 1953 to 1967 and the Conservation of Wild Creatures and Wild Plants Act 1975 to prohibit certain methods of killing or taking wild animals
- amended the law relating to the protection of certain mammals
- restricted the introduction of certain animals and plants into the countryside
- amended the Endangered Species (Import and Export) Act 1976
- amended the law relating to nature conservation
- made certain provisions respecting the Countryside Commission
- amended the law relating to public rights of way

Part one of the Act deals with the protection of wild birds, their nests and eggs, whilst making provision for authorised persons to control birds in the interests of public health or air safety, to prevent the spread of disease or to protect crops, timber or fisheries. This section also protects certain wild animals and prohibits some methods of killing or taking wild animals. An important clause makes it an offence to pick, uproot or destroy any wild plants set out in the schedule unless authority has been given to do so.

Part two of the Act deals with nature conservation and sets out the procedure for dealing with Sites of Special Scientific Interest, National Nature Reserves and Marine Nature Reserves. It also covers management agreements with owners and occupiers of land and the duties of Ministers of Agriculture with respect to the countryside. Water authorities must further the conservation and enhancement of natural beauty and the conservation of flora, fauna and geological or physiographical features of special interest when carrying out their functions.

Part three deals with public rights of way and specifies that a definitive map shall be kept and that this must be under continuous review. It also prohibits the keeping of bulls in fields crossed by a right of way unless the bull is under ten months old, is not of a recognised dairy breed and is running with cows and heifers. Part three also covers various miscellaneous and general provisions.

The Act very largely applies only to those farmers who have on their land a Site of Special Scientific Interest where legal obligations are laid on the occupier, who will receive help in carrying them out from the Nature Conservancy Council for England (English Nature), the Nature Conservancy Council for Scotland or the Countryside Council for Wales. On land outside SSSIs, that is, the vast majority of the countryside, the obligations of the Wildlife and Countryside Act lie not so much in the Act itself as in the emphasis placed on its moral implications in the debates which went on, during the passage of the bill, in both Houses of Parliament and in the press. These identified a very large body of opinion which wants to see the countryside looked after and managed in a more positive way. This places a moral obligation on farmers and landowners who are responsible for 80 per cent of the countryside. It is an obligation which is far more difficult to follow than one with legal implications because there will always be those who will fail to see what needs to be done or who will seek short term gains without regard for the long term effects.

Sites of Special Scientific Interest

Under sections 28 to 38 of the Wildlife and Countryside Act 1981, Sites of Special Scientific Interest are areas of land or water identified as being of outstanding value for their wildlife or geology by the Nature Conservancy Council for England, the Nature Conservancy Council for Scotland or the Countryside Council for Wales. Whilst there are other areas important for nature conservation, SSSIs are exceptional and many are of international importance.

SSSIs are in trust for future generations, and the management of such areas in the interests of wildlife can be a source of pride and pleasure to the occupier. Such management may place restrictions on certain activities and the relevant council may offer an appropriate payment if loss of income is incurred through entering into a management agreement. Owners/occupiers may also seek grant aid or loans from the council concerned to conserve and enhance the scientific interest of their land.

The appropriate council and the owner/occupier will agree a management plan for the site, detailing potentially damaging operations which may not be carried out or the extent to which they may be carried out. It is an offence for any person who has been formally notified under section 28 to carry out a listed operation without proper consultation, or to cause or permit anyone else to do so. SSSIs are listed in schedules covering local planning authority areas.

ORGANISATIONS CONCERNED WITH CONSERVATION

The Countryside Commission for England
The Countryside Commission for England has responsibility for preserving and enhancing the natural beauty of the English countryside and its enjoyment by this and future generations. It is an independent body funded by the Government.

Following the publication of *New Agricultural Landscapes* in 1974, the Commission has undertaken a number of activities to promote the conservation of lowland agricultural landscapes. During the past ten years, the Commission has spent £8.5 million on conservation by helping to fund amenity tree planting programmes and other projects aimed at replacing some of the landscape features lost due to changes in agricultural practices.

In addition, special Demonstration Farm and New Agricultural Landscape Projects have been set up to show that agriculture and conservation are compatible and that conservation schemes can be carried out on farms at little expense. The Commission helps to support the Farm Conservation Advisers, now appointed on a county basis (see FWAG and the Farming and Wildlife Trust (FWT)).

The Countryside Commission for England, John Dower House, Crescent Place, Cheltenham, Gloucestershire GL50 3RA

The Countryside Commission for Scotland has similar responsibilities to that for England but after April 1992 the Commission will merge with the Nature Conservancy Council for Scotland to form a new agency, Scottish Natural Heritage.

The Countryside Commission for Scotland, Battleby, Redgorton, Perth PH1 3EW

The Countryside Council for Wales was formed on April 1st 1991 by amalgamating the Nature Conservancy Council in Wales with the Countryside Commission's Welsh Division.

The Countryside Council for Wales, Plas Penrhos, Ffordd Penrhos, Bangor, Gwynedd, LL57 2LQ

The Nature Conservancy Council (NCC)

Prior to 1991 the Nature Conservancy Council was the Government agency responsible for nature conservation in England, Scotland and Wales. Following the 1990 Environmental Protection Act the Nature Conservancy Council was divided into the Nature Conservancy Council for England (English Nature), the Nature Conservancy Council for Scotland and the Countryside Council for Wales (which also incorporates the Countryside Commission Welsh Division). Each of these bodies will promote nature conservation through the provision of advice, the management of nature reserves and the implementation of legislation relating to wildlife and natural features. They will select and notify Sites of Special Scientific Interest to planning authorities and owners and occupiers, and can offer management agreements, as well as advice, to help farmers and landowners to maintain and enhance this special interest.

The Nature Conservancy Council for England, Northminster House, Northminster, Peterborough PE1 1UA

The Nature Conservancy Council for Scotland, 12 Hope Terrace, Edinburgh EH9 2AS

The Countryside Council for Wales, Plas Penrhos, Ffordd Penrhos, Bangor, Gwynedd LL57 2LQ

The Nature Conservancy Council's successor agencies have set up a Joint Nature Conservation Committee (JNCC) to coordinate their work and to discharge their international conservation responsibilities on behalf of the UK.

The Farming and Wildlife Trust and the Farming and Wildlife Advisory Groups (FWAGs)

The Farming and Wildlife Trust is an independent organisation which has the committed support of the leading farming, landowning and conservation bodies in England, Scotland, Wales and Northern Ireland. The Trust supports the continuing development of the FWAGs throughout the United Kingdom.

The FWAGs aim to further understanding and cooperation between farmers and conservation interests in a practical way. Over the last twenty years, while conflicts between farming and conservation have made headlines, the FWAG movement has gained steadily in strength and influence as an agent through which the farming and landowning organisations, the advisory and educational service, the Countryside Commissions, the Nature Conservancy Council and the leading conservation organisations, can act. Together they explore how wildlife and landscape conservation can be integrated with modern agriculture in a practical way on all farms throughout the country.

There are over sixty groups throughout England, Scotland, Wales and Northern Ireland, bringing together farming and conservation interests with the particular aim of establishing a system by which farmers can receive advice on conservation. Farm walks, demonstrations and seminars have resulted in an ever-increasing number of farmers calling on FWAGs for advice. Over forty FWAGs have a full-time Farm Conservation Adviser, and these advisers have visited more than 20,000 farms throughout the UK during the past six years. Independent studies show that these advisers enjoy the strong support of the farming community and for every £1 spent on providing advice, farmers spent £3 to £4 on conservation activities on farms. There is a remarkably high take-up of advice given and, in addition to the direct spending on conservation, much advice relates to changes in farm practice which cannot be quantified, but which probably has even more far-reaching effects.

The FWT and FWAGs are committed to encouraging and helping farmers and landowners to exercise their responsibility for the conservation of wildlife and landscape on their land, while continuing to farm efficiently and productively.

The FWT also has an educational role with County Agricultural Colleges and the Agricultural Training Board.

The Farming and Wildlife Trust, National Agricultural Centre, Stoneleigh Park, Warwickshire CV8 2RX

Index

Figures in bold type indicate page numbers in the colour section between pages 116 and 117.

141

About the Authors

A Cambridge graduate, Mike Soper lectured at Reading University before joining the Sudan Agricultural Service. From 1946-81, he lectured at Oxford, where he was Director of the University Farm and a Fellow of Christ Church. He organised the Oxford Farming Conference for 30 years, was Chief Assessor to the National Certificate in Agriculture Examinations Board from 1962-87, and chaired the City and Guilds Agriculture Committee for 15 years, being elected to the Honorary Fellowship of the Institute in 1988 for services to agricultural education. He served as Chairman of the Association of Agriculture Executive Committee from 1984-90. A Foundation Fellow of Royal Agricultural Societies, he now serves as Secretary of the English Panel of the Fellowship (FRAgS) Scheme.

Eric Carter is a graduate of Reading University, and his distinguished career in the Agricultural Development and Advisory Service culminated in his appointment as Deputy Director General in 1975. In 1981 he became National Adviser to the Farming and Wildlife Advisory Group. A Council member of the RASE, he edits its Journal and is on the Governing Body of the AFRC Institute for Grassland and Environmental Research. He is a visiting lecturer at the Universities of Reading and Nottingham, a member of the advisory committee of the Centre for Agricultural Strategy and convener of the Standing Conference on Countryside Sports.

FARMING PRESS BOOKS

Listed below are a number of the agricultural and veterinary books published by Farming Press. For more information or for a free illustrated book list please contact:

**Farming Press Books, 4 Friars Courtyard
30-32 Princes Street, Ipswich IP1 1RJ, United Kingdom
Telephone (0473) 241122**

Pearls in the Landscape   CHRIS PROBERT
The creation, construction, restoration and maintenance of farm and garden ponds for wildlife and countryside amenity.

Farm Woodland Management
BLYTH, EVANS, MUTCH, SIDWELL
Covers the full range of woodland size from hedgerow to plantation with the emphasis on economic benefits allied to conservation. Second edition.

The Horse in Husbandry   JONATHAN BROWN
Photographs of horses working on farms from 1890 to 1950, with an account of how they were managed.

A Way of Life   H GLYN JONES AND BARBARA COLLINS
A complete guide to sheepdog training, handling and trialling, in which Glyn Jones' life is presented as an integral part of his tested and proven methods.

The Blue Riband of the Heather   E B CARPENTER
A pictorial cavalcade of International Sheep Dog Society Supreme Champions from 1906, including information on pedigrees, awards and distinguished families of handlers.

Showman Shepherd   DAVID TURNER
A guide to all the facts of showing sheep.

Tractors Since 1889   MICHAEL WILLIAMS
An overview of the main developments in farm tractors from their stationary steam engine origins to the potential for satellite navigation. Illustrated with colour and black-and-white photographs.

Farming Press Books is part of the Morgan-Grampian Farming Press Group which publishes a range of farming magazines: *Arable Farming, Dairy Farmer, Farming News, Livestock Farming, Pig Farming, What's New in Farming.* For a specimen copy of any of these please contact the address above.